The One-Day Expert Series

IMPLEMENTING STANDARDIZED WORK

Writing Standardized Work Forms

The One-Day Expert Series

Series Editor
Alain Patchong

PUBLISHED

Implementing Standardized Work: Writing Standardized Work Forms
Alain Patchong

Implementing Standardized Work: Measuring Operators' Performance
Alain Patchong

FORTHCOMING

Implementing Standardized Work: Process Improvement
Alain Patchong

The One-Day Expert Series

IMPLEMENTING STANDARDIZED WORK

Writing Standardized Work Forms

Alain Patchong

CRC Press
Taylor & Francis Group
Boca Raton London New York

CRC Press is an imprint of the
Taylor & Francis Group, an **informa** business

A PRODUCTIVITY PRESS BOOK

CRC Press
Taylor & Francis Group
6000 Broken Sound Parkway NW, Suite 300
Boca Raton, FL 33487-2742

© 2013 by Taylor & Francis Group, LLC
CRC Press is an imprint of Taylor & Francis Group, an Informa business

No claim to original U.S. Government works

Printed on acid-free paper
Version Date: 20121218

International Standard Book Number: 978-1-4665-6354-4 (Paperback)

This book contains information obtained from authentic and highly regarded sources. Reasonable efforts have been made to publish reliable data and information, but the author and publisher cannot assume responsibility for the validity of all materials or the consequences of their use. The authors and publishers have attempted to trace the copyright holders of all material reproduced in this publication and apologize to copyright holders if permission to publish in this form has not been obtained. If any copyright material has not been acknowledged please write and let us know so we may rectify in any future reprint.

Except as permitted under U.S. Copyright Law, no part of this book may be reprinted, reproduced, transmitted, or utilized in any form by any electronic, mechanical, or other means, now known or hereafter invented, including photocopying, microfilming, and recording, or in any information storage or retrieval system, without written permission from the publishers.

For permission to photocopy or use material electronically from this work, please access www.copyright.com (http://www.copyright.com/) or contact the Copyright Clearance Center, Inc. (CCC), 222 Rosewood Drive, Danvers, MA 01923, 978-750-8400. CCC is a not-for-profit organization that provides licenses and registration for a variety of users. For organizations that have been granted a photocopy license by the CCC, a separate system of payment has been arranged.

Trademark Notice: Product or corporate names may be trademarks or registered trademarks, and are used only for identification and explanation without intent to infringe.

Library of Congress Cataloging-in-Publication Data

Patchong, Alain.
    Implementing standardized work : writing standardized work forms / Alain Patchong.
         pages cm. -- (The one day expert)
    Includes bibliographical references and index.
    ISBN 978-1-4665-6354-4
    1. Performance standards. 2. Labor productivity. 3. Standardization. I. Title.

HF5549.5.P35P345 2013
651'.29--dc23   2012045711

Visit the Taylor & Francis Web site at
http://www.taylorandfrancis.com

and the CRC Press Web site at
http://www.crcpress.com

Contents

Acknowledgments vii
Preface ix

Chapter 1 Introduction 1

Chapter 2 Training Day 3

Chapter 3 "We Already Have Standards!" 9

Chapter 4 On Which Tasks Can Standardized Work Forms Be Written? 25

Chapter 5 Collecting Data 31
    Task One: Group Forming 31
    Task Two: Team Organization 32
    Task Three: Facilitator Gives Job Instructions for Operators 34
    Task Four: Teams List the Detailed Steps on Flipcharts 35
    Task Five: Facilitator and All Teams Determine Measurement Steps and Measurement Points 36
    Task Six: Each Team to Perform 20 Reps 42

Chapter 6 Introducing the Four Standardized Work Documents ... 49

Chapter 7 Getting Ready 53
    The First Element of Standardized Work Forms: Takt Time 54
    The Second Element of Standardized Work Forms: Job Sequence 58
    The Third Element of Standardized Work Forms: Standardized Work in Process (SWIP) 58
    The Fourth Element of Standardized Work Forms: Key Points 60

Chapter 8 Writing the Forms 67

Initiation to the Yamazumihyo or Process Analysis Chart 67
Writing the Three Remaining Standardized Work Forms 78

Chapter 9 Shop Floor Application 83

Chapter 10 Takeaway 91

Index 95
About the Author 101

Acknowledgments

The One-Day Expert series is the direct consequence of my previous work at Goodyear. I owe thanks to several of my former colleagues who provided me with valuable remarks and comments.

I am thankful to Mike Kipe who was part of the team I formed to deploy "Standardized Work" in Goodyear plants.

Throughout the writing process, I received numerous ideas and suggestions from Dariusz Przybyslawski. Besides being part of the Standardized Work deployment team at Goodyear, his help was instrumental in structuring and tuning The One-Day Expert series as well as establishing the proof for the concept of such a project.

I am very grateful to two other former colleagues at Goodyear who, at a very early stage, believed in Standardized Work as presented in this book, and gave me the opportunity to try it on the shop floor: François Delé and Markus Wachter.

I am obliged to Alain Prioul, Xavier Oliveira, Pierre-Antoine Rappenne, and Philip Robinson, who read drafts and offered valuable suggestions for improvement. I am also very thankful to the editorial staff of Taylor & Francis, especially Jay Margolis, for their wonderful work in improving the readability of the initial text.

I express my gratitude to all of my colleagues at Faurecia who work with me in testing ideas and actions, and critique or support my thoughts.

Finally, and most especially, I would like to give my special thanks to my wife Patricia, my son Elykia, and my daughter Anya for their unrelenting support and patience.

Preface

The One-Day Expert series* presents subjects in the simplest way, while maintaining the substance of the matter. This series allows anyone to acquire quick expertise in a subject in less than a day. That means reading the book, understanding the practical description given in the book, and applying it right away, in only one day. To focus on the quintessential knowledge, each The One-Day Expert book addresses only one topic and presents it through a streamlined, simple, narrative story. Clear and straightforward examples are used throughout each book to ease understanding, and thereafter, application of the subject.

---

* For more details on the series, please go to The OneDayExpert.com.

1
Introduction

"Standardized Work deployment is the DNA coding of operational excellence." The first step of Standardized Work deployment consists of capturing the current state. This book is the second on Standardized Work deployment, and is dedicated to additional tools that help capture the current state.[*]

An operation can be standardized only if it is repeatable. Therefore, the most important requirement for a real application of Standardized Work is minimum stability in the process. The stability level of a process can be estimated by performing Process Analysis. Process Analysis has two outputs. First, it helps determine the real capacity of the process. This means the maximum and the average production that the process yields, which are then compared to customer demand. Second, it shows the main cause for capacity losses and especially the contributions of the main sources of variation: deterministic causes (e.g., material and code change) and stochastic causes (e.g., machine breakdown, quality problems, material-related or personnel-related variability). All other causes need to be addressed, especially when they are important contributors, before personnel-related variation. In this sense, Process Analysis helps set the right priority for action. In this book, we present a tool to perform Process Analysis. It is a multistage bar called *Yamazumihyo*[†] in several Japanese companies.

Besides Process Analysis, other Standardized Work forms are addressed in this book: the Standardized Work Chart, the Standardized Work

---

[*] The first book is dedicated to Operators' Performance Measurement.
[†] This book uses the spelling "Yamazumihyo." Other variations including Yamazumi hyo and Yamazumihyou exist. "Yamazumihyo" is actually the concatenation of three words in Japanese: Yama, which means mountain; Zumi, which means accumulate; and Hyo, which means chart. Like many Japanese companies, we use the word "Yamazumihyo" to characterize the Process Analysis chart. However, it may be used to designate something else in some other Japanese companies.

Combination Table, and Operator Work Instructions. These three documents focus on work methods and conditions. They are built around key elements of operator's work: Takt time,* Job Sequence,† Standardized Work in Process, and Key Points.

In this series, The One-Day Expert Series dedicated to Standardized Work, Thomas, a young, high-potential plant manager in an industrial group, is reassigned to another plant, which is losing money. Previous plant managers have tried several initiatives with, at best, limited results. His urgent mission, which sounds like the EMEA‡ senior management's last card, is very simple: turn the plant around. The morale in the plant is very low; the staff is equally pessimistic about the plant's future and distrustful of senior management. Time is running out; company headquarters needs concrete results and has become impatient. To face these challenges, Thomas has decided to use Standardized Work deployment to achieve quick and visible results while rebuilding a real team. To this end, he requested the support of Daniel Smith, the Industrial Engineering Manager for EMEA. Daniel had been with the company for only a couple of years, after previous experience in the automotive industry. Building on his previous experience, he recently designed and launched a Standardized Work initiative and is looking to prove the real power of Standardized Work by deploying it in several plants.

The series of books dedicated to Standardized Work improvement recounts, step by step, Thomas's deployment of Standardized Work with Daniel's support. The first book of this series shows the initial steps Thomas took to assess the plant's current situation through measurement of operators' performance. In this second book, we will learn about the next steps of this assessment, which consist of writing Standardized Work forms to help see both variability and waste.

---

* Takt time, also called the voice of the customer, is the time between two consecutive parts representing customer requirement. This is the result of the production time divided by the corresponding customer requirement.
† Also called Work Sequence.
‡ EMEA stands for Europe, Middle East, and Africa.

2
Training Day

On this 68°F day in June, Daniel had a very tough wakeup after spending the previous evening socializing. He had accepted an invitation to have dinner with Steve, the plant Industrial Engineering (IE) manager and a few other employees. The dinner lasted late into the night, which explained why he felt a little tired. Steve was a pure IE guy. He had been with the company for 20 years now. He got an internship with Home Appliances, Inc. during his last year at University, while completing his master's degree in Industrial Engineering. He made a very good impression during that period, which landed him a job at the company. Steve started as an IE Coordinator. The IE department he joined then had 10 people and an assistant. At that time, he was a young and ambitious employee that many colleagues respected for his professionalism. Steve's co-workers also liked his high competence in mastery of numeric and analytic matters. If there were any complex issues around, then they belonged to Steve. As most people used to say, "If Steve can't solve it, then nobody can." Despite his glittering notoriety, Steve did not want to be indispensable. Therefore, he had developed training materials and easy-to-use tools, and designed and taught some simple rules to help his colleagues solve problems without his support. His natural intellectual authority had prompted some to call him teasingly the "Guru" or "the big Guru." Besides his recognized professionalism, he was overwhelmingly hailed as a very affable and easygoing person. No surprise that many colleagues befriended him.

Steve worked his way to the top of the IE department. But the IE department he inherited had lost more than half of its staffers. Over the years, perception of the IE department by regional senior management had evolved. The department once considered as the problem-solver-and-important-number-reservoir had morphed into a unit perceived merely as a contributor to indirect compensation costs. As Steve complained,

"We had to let people go and continue to do the same thing, or even more, with fewer and fewer people."

Several reports were issued by the IE department on a predefined frequency basis. From the monthly staffing report to the weekly OEE[*] document, Steve and his team authored more than 20 documents on a periodic basis. Of course, not all of them were useful. He was not even sure how many of their recipients bothered to have a look. Nevertheless, issuing those documents was definitely part of his job. This is how he learned to do the IE job. It was, as he called it, their "core business activity." He would voluntarily recognize that things could be improved, but he was simply not ready for that. Moreover, he feared that by reducing the amount of work he would be exposed to a request for even more staff reductions. That would mean fewer people to manage, which he perceived as a loss of power and even a demotion. The rule was simple in this plant, and everybody knew it: "A manager's power is proportional to the number of people she or he manages."

In the course of the dinner, Steve also signaled his unease about the increasing importance that was being given to Excellence System[†] people. As he put it:

> Those guys are doing some of the things we used to do. Also, they tend to oversimplify some subjects. Most of them believe that in case of a problem they just have to go on the shop floor, what they call Gemba,[‡] perform a workshop, and "voilà" everything is solved. They are cannibalizing us. More broadly, they are putting Lean everywhere thereby getting the whole credit for any work done. We work, they get the credit … how dare they do that? This situation is de-motivating my team. Daniel, we hugely need your help.

These words came as no surprise to Daniel. He had heard them before while visiting other plants in the EMEA region. Most IE people felt like they were being left behind at the benefit of Excellence System. To say the least, their morale was not at its height. By all accounts, the IE community was facing a somewhat gloomy prospect. Industrial Engineering departments across the region had found themselves in an unyieldingly perilous fight for relevance, if not survival.

---

[*] Overall Equipment Effectiveness.
[†] Excellence System was the name given to the companywide Lean deployment initiative.
[‡] This Japanese word means shop floor, where real things happen; where the value is added.

Daniel took the time to listen to Steve and gave him some assurance regarding the evolution of IE positions in the company. He underscored that the senior management had no intention of eliminating IE or diluting it into the Excellence System department. These were two functions that would coexist, and therefore needed to work together to solve the plant's problems.

Wednesday was the third day of the training.* The weather that morning, despite the proximity of summer, looked very autumnal due to the dark sky and unstoppable rain. As always, this morning Daniel arrived 30 minutes before the start of the training, notwithstanding his short night. He needed those 30 minutes to write down quick highlights of the previous day's activities and lay out the Key Points regarding the training day. These were part of Daniel's own standard for the organization of the training.

Thomas, who had decided to change his week's schedule to attend the training,† came a few minutes after Daniel. He took the free time before the start of the training to chat a little bit with the trainees. He wanted to make sure that the training was well received by everyone. He also had a quick discussion with Daniel to get a flavor of the day's agenda.

This Wednesday the room had more participants. Thomas explained the reason for that:

> I found the module on Operators' Performance Measurement to be very productive. I thought that it would help the deployment if some of the key managers of this plant could follow the rest of the training. I therefore requested, if they could free up their calendar, that each of them attend from today to the end of week. For this module of the training, focusing on the second tool of Standardized Work, seven managers reporting directly to me are going to be able to attend. I know it was not easy for them to cancel or move some of their very important meetings, especially the ones scheduled long ago. I do appreciate their effort and thank them for that. I am sure that at the end of the week they will have no regrets.

Daniel, who got word about those changes from Thomas on Tuesday evening, made the necessary logistics changes to welcome the new participants. Some adaptations were definitely needed as the group of trainees went from the initial eight people scheduled to sixteen people on

---

* The first two days are the subject of the first book: Implementing Standardized Work: Measuring Operators' Performance.
† Episode recounted in the first book: Implementing Standardized Work: Measuring Operators' Performance.

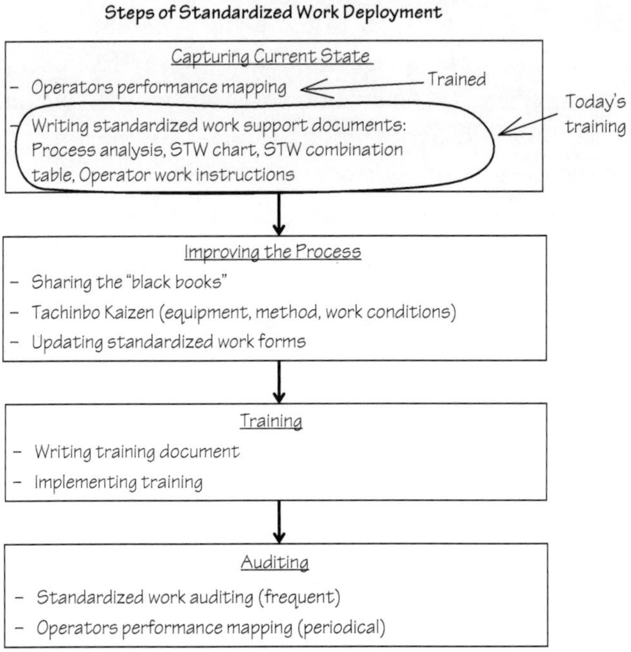

Steps of Standardized Work Deployment

```
Capturing Current State
- Operators performance mapping  ← Trained
- Writing standardized work support documents:    ← Today's training
  Process analysis, STW chart, STW combination
  table, Operator work instructions
         ↓
Improving the Process
- Sharing the "black books"
- Tachinbo Kaizen (equipment, method, work conditions)
- Updating standardized work forms
         ↓
Training
- Writing training document
- Implementing training
         ↓
Auditing
- Standardized work auditing (frequent)
- Operators performance mapping (periodical)
```

FIGURE 2.1
Chart of the steps of Standardized Work deployment.

Wednesday morning. This was a minor inconvenience to Daniel in light of his delightedness at the success of his previous training.

At 8:00, the scheduled time for the beginning of the training, a few people were still missing. Daniel thought this was probably due to early rain that disturbed the traffic. "This is an exact illustration of the impact of variability," he joked. "The people who are late are probably the ones who had not enough or no buffer time in their commuting process. This is exactly what happens in the plant. If you cannot conquer or absorb variability, your customer will wait, and you will lose money."

Daniel, who against all odds looked relaxed and loose, went on:

> Well, it is 8:10, there is only one participant missing; I suggest we start. Before we begin with the materials I have prepared for you today, I propose we make a quick summary of what we have accomplished so far.

Daniel walked to the flipchart and continued: "Let us pull back the chart we discussed yesterday (Figure 2.1). Now, can you tell me what are your main takeaways from this chart?"

After a few seconds of silence, one of the participants raised his hand and commented:

> The Key Point I took from this chart, and more generally from the past two days, is that Standardized Work should be deployed as a system composed of four steps: Capturing the Current State, Improving the Process, Training, and Auditing. Each of these steps, deployed individually, will yield very few results that may fade very quickly. A system approach is the fabric that strengthens the deployment and amplifies the magnitude of the results.

"Very good summary of the big picture," Daniel commented. "Now, if we talk about the first step—Capturing the Current State—what can you say about that?"

At this time, another participant jumped in and added:

> Over the last two days, we have learned how to measure operators' performance. I have to say that I was especially surprised by the role of variability and its hidden cost. I also learned that the operator visibly working at the highest pace was not automatically the best performer. Above all, the performance measurement method you taught us looks very simple … And it is so powerful. I look forward to its application in the whole plant.

"That's very good. Anything else to add … ?" After a few seconds of silence Daniel continued:

> Today we will focus on several tools that will help capture the current state. Those tools can be classified in two categories. The first category is comprised of the process analysis chart called Yamazumihyo in many Japanese companies. I often use the term Yama for short. This tool is used to assess the potential capacity of your process, the losses, and the sources of those losses. The second category includes three forms as you may see on the chart we just reviewed (Figure 2.1). Those three forms are Standardized Work Chart, Standardized Work Combination Table, and Operator Work Instructions. Today we will learn how to write those four documents. Before we proceed, it's time to quickly mention a caveat. Some of you who have some knowledge of Standardized Work may not be familiar with some of the tools I just mentioned. Yes, there are some small changes versus Standardized Work, as you may see in the literature, or as it is applied at Toyota. I will come back to those discrepancies and explain the reasons.

3
"We Already Have Standards!"

After this quick introduction, Daniel wanted to address right away a lingering question about new standard introduction.

> Before I continue, let me address upfront questions and remarks most people ask me when I talk about writing Standardized Work forms. This is what I hear very often: "We already have standards! What is different?" and "Do we really need this?" To answer these questions and get you ready for the training day, I have prepared two warm-up quizzes for you.

Daniel went to the table and grasped a pile of papers that he handed out to each participant.

> Of course, most of the plants already have some standards. The first two papers in the pile I just handed out are an example of a standard I got from the Industrial Engineering department of one of our German plants that produces bladders for some of our home appliances (Figure 3.1 and Figure 3.2). I guess you've got similar ones here, correct?

The Industrial Engineering manager nodded. Then Daniel continued:

> Well … for sure you already have standards. Now, the question is what is the difference between those standards and the one we will discuss today? I suggest we come back to this question after the warm-up quizzes I have readied for you. Before we move further, let me tell you that the original version of this two-page IE standard had three full pages. The version you have in your hands is a lighter one (Figure 3.1 and Figure 3.2). I have removed some unnecessary data from the original form to obtain a more readable document for you.

Time and Motion Study—Page 1/2

Time Table	Machine Cycle	PW* Time	Manual Time	Motion Freq.	Motion Time	PW Time	Total PW Time
	Component 1 Application						
0.000		0.040					
0.010	– Component 1 server positioning						
0.020	and spot drum for application						
0.030							
0.040	– Fix Component 1 on drum and	0.120	– Operator stays at drum to			0.120	
0.050	application (1) turn		monitor Component 1 application				
0.060	– Partial withdrawal of Component 1	0.500					
0.070	server and spot for splice						
0.080							
0.090		0.210	Idle time:	0.090		0.120	0.210
0.100							
0.110			– Prepare Component 1 splice	1-1	0.200	0.200	0.410
0.120							
0.130			– Step out of safety area and push	1-1	0.030	0.030	0.440
0.140			splicer cycle start button				
0.150							
0.160	**Splicer Cycle**						
0.170	– Splicer positioning	0.060	– Prepare (1) Component 4 for	1-1	0.120	0.120	
0.180			Component 3 splice				
0.190	– Stitch Component 1 splice	0.097					
0.200							
0.210	– Splicer withdrawal and	0.040					
0.220	Component 2 server forward						
0.230	motion						
0.240		0.197	Idle time:	0.077		0.120	0.637
0.250							
0.260	**Component 2 Automatic Application**						
0.270	– Splice for Component 2	0.020	–Grasp (2) Component 2 and fix	2-1	0.030	0.060	
0.280			it on drum				
0.290			– Start Component 2 application	1-1	0.040	0.040	
0.300			and grasp hot knife				
0.310	– Component 2 application, no	0.060				0.020	
0.320	inching						
0.330			– Cut (2) Component 2 using hot	2-1	0.085	0.170	
0.340			knife, push cycle advancement				
0.350							
0.360	– Component 2 server withdrawal	0.040	– Make (2) Component 2 splices	2-1	0.070	0.140	
0.370			and push cycle advancement				
0.380							
0.390						0.430	1.067
0.400							
0.410	**Component 3 Application**						
0.420	– Component 3 server forward	0.050	– During Component 3 server	1-1	0.050	0.050	
0.430	motion and spot drum for		advancement grasp Component 3				
0.440	Component 3 application		– Fix Component 3 on shelf and	1-1	0.040	0.040	
0.450			start application				
0.460	– Component 3 application (1)	0.070				0.070	
0.470	turn no inching						
0.480			– Cut Component 3, aside with	1-1	0.130	0.130	
0.490			knife				
0.500	– Component 3 server withdrawal	0.040	– Grasp hand splicer	1-1	0.020	0.020	
0.510							
0.520			– Prepare splice and position	1-1	0.090	0.090	
0.530			hand splicer				
0.540			– Make splice with hand splicer	1-1	0.060	0.060	
0.550							
0.560			– Put aside hand splicer	1-1	0.020	0.020	
0.570							
0.580							
0.590						0.480	1.547
0.600							

*PW = Production Work

FIGURE 3.1
Industrial Engineering standard.

Time and Motion Study—Page 2/2

Time Table	Machine Cycle	PW* Time	Manual Time	Motion Freq.	Motion Time	PW Time	Total PW Time
0.720	**Component 4 Manual Application**						
0.610							
0.620	– Spot for Component 4	0.020	– Grasp prepared Component 4	1–1	0.030	0.030	
0.630							
0.640			– Fix Component 4 on Component 3 splice, rotate drum for manual center stitch	1–1	0.060	0.060	
0.650							
0.660							
0.670			– Step out of safety mat and push button to start Component 4 stitch cycle	1–1	0.030	0.030	
0.680							
0.690							
0.700						0.120	1.667
0.710							
0.720	**Component 3 Stitching**						
0.730							
0.740	– Component 3 stitchers up	0.060	– Load bead loader with (2) preassembled bead and start transfer cage cycle	2–1	0.090	0.180	
0.750							
0.760	– Component 3 stitching	0.310					
0.770							
0.780	– Spot drum and crossrail unit withdrawal	0.040					
0.790							
0.800							
0.810							
0.820			Idle time:	0.230		0.180	2.077
0.830							
0.840	**Finished Good Removal**						
0.850							
0.860	– Finished good transfer cage in (from waiting position)	0.080	– Operator step for finished good removal				2.340
0.870							
0.880	– Suction cup down and aspirate finished good	0.120					
0.890							
0.900	– Finished good transfer cage out to waiting position	0.070					
0.910							
0.920							
0.930		0.270	Idle time:	0.270		0.000	2.347
0.940							
0.950							2.347

– Net building time:					141		2.347
– Miscellaneous machine and material allowances of net building time						2.00%	0.047
– Lost time due to roll changes:							
a) Wire ply				102.4	0.200		0.002
							2.396
No Relief:							**2.396**
– Tools preparation:							
a) Secure tools at the start of shift and aside tools at the end of shift					5.000		
				Bladders/shift	164		0.030
– Standard time:							**2.701**
– Personal time:						5.00%	0.135
– Total time:							**2.836**
– Piecework requirements/shift (7.75 hrs)							**163.95**

*PW = Production Work

FIGURE 3.2
Industrial Engineering standard.

12 • Implementing Standardized Work

Name:

Questions	Answers
1. What is the operator doing (major steps)?	
2. What is the cycle time of this operation?	
3. How much time does the operator spend waiting?	
4. How much time does the operator spend walking?	
5. Do we have the capability to satisfy customer demand?	☐ Yes ☐ No ☐ Unsure
6. What could we do to reduce the cycle time?	
7. Could you use this/these form(s) on the shopfloor to tell if an operator is working according to the standard?	☐ Yes ☐ No ☐ Unsure

FIGURE 3.3
Warm-up quiz on standards.

Now let's move on. Everyone pick up the next paper; I mean the third one (Figure 3.3). On this paper, you will find seven questions regarding the process the IE standard is related to. Now, to make sure that everyone understands, let us go through all those questions one by one. I will read them all, and if you find any ambiguity, please stop me and I will explain. All right … Let's get started: Question 1: What is the operator doing (major steps)? You should answer this question and the following one based on what you see or read on the IE standard. Is that clear?

Most of the trainees acquiesced.

Okay, let us move to the next ones, Question 2: What is the cycle time of this operation? Next question, Question 3: How much of the time does the operator spend waiting? Next one, Question 4: How much time does the operator spend walking? I'm not getting a reaction from you guys. Does that mean everything is clear so far to you?

Daniel got no answer from the participants, who looked very focused.

Next question, Question 5: Do we have the capability to satisfy customer demand? You have the choice between three answers here. Based on the IE standards you should be able to tell if you have clear information regarding customer demand or customer requirement satisfaction. There are three possible answers. Answer 1 is "Yes," the machine has the capability to satisfy customer demand. Answer 2 is "No," the machine does not have the capability to satisfy customer demand. Check the third answer box if you are unsure. Is that clear for everyone?

> *Remember that one of the goals of the Standardized Work forms is to help us see losses or waste and eliminate them.*
>
> *Also, you need to be able to use them to check if the operator is actually working as defined in the Standardized Work.*

Most of the trainees acquiesced this time.

Now, let us move to Question 6, What could we do to reduce the cycle time? This is an open question. Please remember that one of the goals of the Standardized Work forms is to help us see losses or waste and eliminate them. Therefore, I am asking you what improvement actions can you propose based on these forms.

Daniel paused, looked around, and then proceeded:

If there are no questions, I suggest we go to the last one. Question 7: Could you use these forms on the shop floor to tell if an operator is working according to the standard? Remember, this is also one of the goals of Standardized Work forms. You need to be able to use them to check if the operator is actually working as defined in the Standardized Work. The question here is very simple. With those documents in hand, standing before an operator, will you be able to say whether he or she is working according to the standard? It seems like all the questions are clear. From this moment, you have five minutes to answer them. Go! I just started the clock.

Daniel used the five minutes to move from table to table. This was part of his "ritual." The answers he saw were usual as well. "No surprise once again," he murmured. Most of the trainees were struggling to find the answers to the questions. He remembered that he struggled the same way the first time he tried to read these standards.

"Time's up!" Daniel shouted.

The five minutes given have now expired. Let's see what you've got. First of all, here is a quick comment. When I walked around, I noticed that many of you were having a hard time finding the right answers. I should say this is somewhat normal. I struggled the same way the first time I tried to answer those questions. I hope this gives you a little bit of a comfort and reassures you. Because I was not able to answer those questions myself, I asked the

IE guy who wrote these standards to give me the answers that I am going to share with you very shortly. I also asked the same guy to let me know how much time it took him to get to those answers. His response: "twenty minutes." Now, here are the answers.

Daniel unveiled a chart he had stuck on the wall and hid that morning while preparing the training.

Let us consider the answers to the questions I wrote on this chart (Figure 3.4). We start with the first question's answer, which, by the way, is pretty obvious. The steps are basically all the titles appearing in the IE standard. Those titles, as you may see on this chart, are process or machine-centered, as opposed to an operator-centered approach. This is a common point in most of the IE standards I have run into so far. These titles are the best answer to the question. They can be considered as major steps, albeit they do not describe clearly what the operator is doing.

After this quick digression, Daniel refocused diligently:

Most of you have probably got this one right. Next question, the cycle time … What is the cycle time of the operation? It looks like a simple question, but a lot of people may find it confusing because IE standard gives the net cycle time, the cycle time with allowance, the cycle time with changes, the cycle time with tool preparation, and the cycle time with personal time. It might therefore be difficult to figure out which number is requested. The IE answer here is what is called the net cycle time in the IE standard, which is 2.347 minutes. The third question about the waiting time of the operator is one of the most complicated to answer. To find the answer you would need to do some calculation as shown on the chart. Some of the numbers appearing in that sum are deduced from other numbers. I mean, this is not really simple … the number is 0.80 minutes. Besides that, a word, "shaping," you have not seen before appears in the answer. You really have to know more about the process than what is shown in the standard to figure it out. I am quite sure that none of you got this one right.

Let us look at the next question, the fourth one. The walk time is not easy to find either. Just as with the wait time, you need to do some arithmetic. The answer offered by our IE colleague is 0.15 minutes. I am sure that none of you got this one right either. Next question, the fifth, … Okay unless you've got some hidden talent at guessing, no information is actually given regarding what is the expectation of the customer. We do not know what the customer demand is. Consequently, the answer to this question should be "Unsure." Now, what can we say about the sixth question?

Questions	Answers
1. What is the operator doing (major steps)?	Component 1 application splicer cycle Component 2 automatic application Component 3 application Component 4 manual application Component 3 stitching cycle Finished good removal
2. What is the cycle time of this operation?	net time = 2.347 min (141 seconds)
3. How much time does the operator spend waiting?	+ 0.21 (monitoring Component 1 application) + 0.07 (splicer cycle) + 0.23 (stitching cycle) + 0.283 (shaping) + 0.007 (Finished good pick-up cycle) = 0.80 min (48 sec)
4. How much time does the operator spend walking?	+ 0.03 (step out safety mat to prepare gum strip) + 0.03 (step out safety mat to prepare beads) + 0.03 (to load beads) + 0.03 (step remove finished good) + 0.03 (return to machine) = 0.15 min (9 sec)
5. Do we have the capability to satisfy customer demand?	☐ Yes ☐ No ☒ Unsure
6. What could we do to reduce the cycle time?	-Reduce all automatic application cycle times. -Transfer some task to another machine (improvement resulting from a recent Six Sigma project).
7. Could you use this/these form(s) on the shopfloor to tell if an operator is working according to the standard?	☐ Yes ☒ No ☐ Unsure

*Answers given by IE Engineer Bladder plant, Germany.
Estimated answering time: 20 minutes.

FIGURE 3.4
Answers to questions on current IE standard.

Thomas reacted, "I laid down a few ideas, but frankly I was not sure about the answer. I guess this is the case for several people here." A handful of people nodded in a sign of approbation. Daniel asked the other trainees to share their improvement proposals. They were very diverse and few of them were supported by data pulled from the IE standard. Then Daniel continued:

> I'll say that there is no right or wrong answer, per say, to this question. As I said previously, a standard is one, if not the most, important tool in any improvement initiative. Therefore, by reading it you should be able to figure out what to improve in a process. What I can see is that the ideas for improvement did not pop off the page. The rule I always use here is that no rational proposals, I mean the ones that are not based on data, will be left unconsidered. I will be generous; give yourself the mark on this question if you have provided at least one acceptable proposal. Should we move to the last question now?

The audience nodded.

> Well, as previously, there is no right or wrong answer to this question either. However, based on all our previous discussions, it seems very unlikely that you will answer "Yes." The right answer is "No." Maybe, except for Steve, the IE manager, I guess very few people in this room are able to check if an operator is following the standardized way or doing the work based on the IE standard document in your hands.

Thomas, who was eager to summarize the lesson as always, asked, "Daniel, what is the takeaway here?"

> Thomas, I suggest we try to answer the same questions based on Standardized Work forms this time. Afterward, we can discuss the outcome. Please pick up the next two papers from the pile I handed out to you (Figure 3.5 and Figure 3.6). Please be aware that a different group of people in a workshop realized those forms; therefore, do not expect to find the same numbers you saw in the IE standard papers. Do not try to copy the numbers from the previous IE standard; nothing you have seen in the previous standard will help you. The questions are the same as previously and you have the same time to answer them. Is everyone clear?

The audience acquiesced. "Let's get started," Daniel launched.

After five minutes, Daniel stopped the exercise and shared the answers with the group (Figure 3.7). He noticed that, unlike previously, several

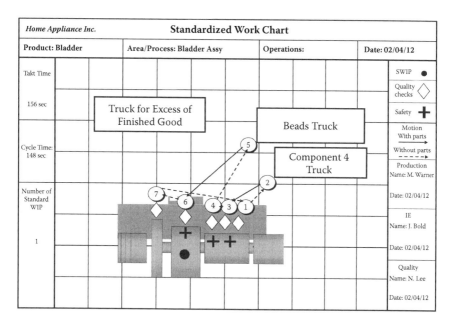

FIGURE 3.5
Standardized Work Chart.

people had completed the quiz. The group went through all the questions and displayed the results of the two quizzes on the same chart (Figure 3.8). He then wanted to get feedback from the trainees. A question popped up very quickly on the Takt time. Daniel knew from experience that despite the Excellence System training having been officially deployed in all plants in the region, many people were still struggling with some simple concepts. He was therefore not really surprised by this question. He had noticed countless times throughout the Standardized Work training that very few people in plants had heard about the Takt time. Never mind that some of them probably did not know how to compute it correctly. For all those reasons, Daniel had decided to include the calculation of the Takt time in this training module. "This is clear evidence that all the talk about successful deployment of Lean is just sound off," Thomas murmured to himself. He explained that the Takt time represents the time between two parts based on the requirement of the customer. He then completed,

> If your cycle time is above the Takt time, then you will not be able to deliver the quantity of products requested by your customer. On the other hand, if your cycle time is below the Takt time you definitely have the capacity to supply your customer if you can manage your variability correctly. By variability, I mean breakdown, changes, etc. In general, when the cycle time is

18 • *Implementing Standardized Work*

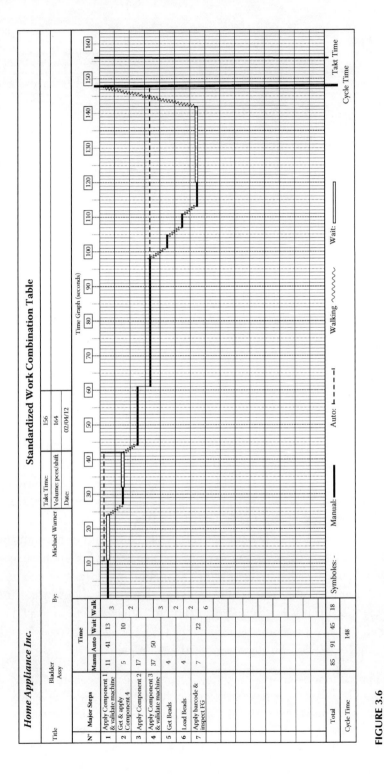

FIGURE 3.6
Standardized Work Combination Table.

Questions	Answers
1. What is the operator doing (major steps)?	1. Apply Component 1 and validate machine 2. Get and apply Component 4 3. Apply Component 2 4. Apply Component 3 and validate machine 5. Get beads 6. Load beads 7. Apply barcode and inspect finished product
2. What is the cycle time of this operation?	148 seconds
3. How much time does the operator spend waiting?	45 seconds
4. How much time does the operator spend walking?	18 seconds
5. Do we have the capability to satisfy customer demand?	☒ Yes ☐ No ☐ Unsure
6. What could we do to reduce the cycle time?	-Seek solutions to increase splice jammer speed so that operator waits less. -Reduce walk time by bringing containers closer. -Check if possible to remove to position 3 before end of machine activity to cover walk.
7. Could you use this/these form(s) on the shopfloor to tell if an operator is working according to the standard?	☒ Yes ☐ No ☐ Unsure

FIGURE 3.7
Answers to questions on Standardized Work forms.

clearly below the Takt time, it does not mean that there is nothing to do. You may simply be using too many resources. In this case, you need to better balance your resources, which will improve your productivity.

He added that he would come back to the Takt time calculation later in the training. In addition, he promised to provide more details using concrete examples.

After this discussion on the Takt time, it was time for Daniel to share trainees' feedback on the differences between standards. To facilitate the discussion, he displayed a 3-column chart, which he built from the

Scores of Warm-up Questions

Score	IE Standard Forms	STW Forms
0		
1	//	
2	→ //////	
3	////	/
4	///	///
5	/	////
6		→ //////
7		////

STW forms are more accessible than IE standard forms: The most frequent (mode) score goes from 2/7 for IE Standard forms to 6/7 for STW forms.

FIGURE 3.8
Distribution of the scores to warm-up quizzes.

previous quiz, showing participants' score distribution (Figure 3.8). In the first column were all score possibilities, ranging from 0 to 7. The second and third columns were respectively titled IE Standard Forms and STW Forms. In front of each first column number was the count of people who got that score: one stroke representing one person. "Well, do you see the difference between those two sets of documents?" he asked.

Thomas reacted, "I think it is pretty clear that the set of Standardized Work documents are way more accessible compared to the IE ones. This is clearly visible when you look at the distribution of correct answers displayed on the chart in front of us." He then stood up, got close to the aforementioned chart, and drew two envelopes in dots. One looked like a valley on the IE standards side and another one looked like a mountain on the Standardized Work side. "I have noticed that the most frequent number of correct answers in the first series of questions was 2. This is what we call the mode, correct?" he said looking toward Daniel, who acquiesced. Then Thomas continued:

> Okay. As I just said, the most frequent number of correct answers was 2 on the IE standards, but we can see that it climbed to 6 when we were asked to answer based on Standardized Work documents. The other thing that those dotted lines are showing is that in the first case, very few people really understood, while in the second case, very few people did not understand.

Daniel thanked Thomas for the summary and asked the rest of the audience to comment. Steve, the plant IE manager, made a remark that Daniel highly expected:

> Daniel, I noticed that all times were expressed in seconds in the Standardized Work document. What is the interest of doing so when it is clear that subtraction and addition are more complicated when the time is expressed in seconds? This is wasteful and not good for us, is it?

"Thank you, Steve, for this excellent question," Daniel replied, with shining eyes. "Most IE people prefer to use minutes in their document because it is very convenient to manipulate them. I mean to add and subtract them. However, I strongly recommend you use seconds …"

At this specific moment Sarah, the plant HR manager, raised her hand and interrupted Daniel. "Daniel, frankly I see no real difference using minutes or seconds, time is time, whatever units you are using do not change anything's real value. Correct?"

"Sarah you are right, units do not change the value of the time," replied Daniel. He then clarified:

> This is not the point. We should remember that the first quality of a Standardized Work form is that it should be easy to read, as it is destined for a large audience; not only IE people but managers, group leaders, team leaders, and operators. Everyone should be able to read it and understand it right away. Coming back to the usage of minutes, you have seen many time numbers in the IE standard that are below one minute. Did you really get a sense of the amount of time they represented when you read them? To be really illustrative, here is a question I have for you: If I tell you that I would need 0.167 minutes to move from here to the first row of chairs, what would you say? Am I fast or am I slow?

The hesitant room stayed silent for a few seconds. Then Steve raised his hand and said, "I can tell." Daniel reacted:

> For sure, Steve, I have no doubt that you can tell. However, as you may notice, you are the only person in this room who can tell. Now, let us suppose that I tell you I need 10 seconds to move from here to the first row of chairs. I am sure that everyone here will have a better sense of my pace. In this specific case, you would most probably say that I am a very slow person as the first row of chairs is just two steps away from my current position. So, Steve, telling someone that the time needed for a given task is 0.167 minutes does

convey the same amount of information as saying that it needs 10 seconds. The value is the same but the understanding is different.

Daniel looked toward Steve and noticed that he was nodding to signify his agreement. Sarah looked convinced as well. He then added, "The main feature a Standardized Work document must exhibit is that it should be easy to read to everyone who can read the numbers, whatever his or her literacy level in the language used in the document."

Thomas raised his hand and enquired:

Daniel, you said that this exercise would help us answer the initial question about why we need Standardized Work forms and what it could bring us. I think most of us would agree that Standardized Work forms are clearly different from the existing. As I commented previously, this is clearly visible on our scores to the two quizzes. If I may summarize, first, Standardized Work forms are easy to read and they give the reader a clear understanding of what the operator is doing. Second, as hinted in my previous point, they are centered on the operator's work and his interaction with machines and parts. This is a clear difference from the IE standard, which seems to be more focused on the machine. Third, the Standardized Work forms give a clear view of expectations because you can see how you are performing versus the Takt time. I would like to remember that the reason we are here is to satisfy our customer. How could we achieve that when we do not even know how we are positioned versus customer demand, which is given by the Takt time? Fourth, the Standardized Work forms make it very easy to see the amplitude of some non-value-added activities like walk and waiting. Therefore, they ought to be excellent tools to support improvement at a workstation. Now, here is the question I am dying to ask you Daniel—Standardized Work forms are excellent, but does that mean that we should throw out our current IE standards?

Daniel, who had gotten this question before, had a prepared answer. He knew from his experience that asking people to get rid of existing documents they have been using for years at the very beginning of a deployment could foster some unnecessary roadblocks going forward. Besides that, he also noticed that IE standards often contain additional machine or product data that could be useful to several functions including IE. When it comes to the introduction of new concepts or tools, he always kept in mind the advice he received from his mentor very early in his career:

"Always wage only one battle at a time; target the big cities and the small ones will fall for themselves." So, not surprisingly, Daniel said:

> The point here is not about throwing your current standards in the trash bin; they do have some data that are important to some folks in this plant. However, they are not sufficient if your goal is to continuously improve your plant. You will need more; you will need Standardized Work forms for all the reasons we discussed before. I therefore suggest that you keep your current standards.

Steve, the IE manager, welcomed this answer with a smile that did not go unnoticed by Daniel.

It was time to take a very short break as Daniel signaled, "Please take a five-minute break, guys. When you return, we will move to the simulation module."

> *The main feature a Standardized Work document must exhibit is that it should be easy to read to everyone who can read numbers, whatever his or her literacy level in the language used in the document.*

4

On Which Tasks Can Standardized Work Forms Be Written?

When the group returned from the break, Daniel thought that it would be good to re-pitch the simulation because half of the participants did not attend the first two days.[*] Thereafter, he continued, "Let us share a few key data points that you will need later on." He moved to the flipchart and wrote while explaining:

> First, let us start with the customer need (Figure 4.1). Our customer is requesting 17,280 T-shirts per day. On the production side, we have four stations; we are producing in three shifts of 8 hours. One operator runs each station and is replaced during breaks. It means that we are producing continuously during 24 hours, which makes a total of 86,400 seconds per day. Quick comment regarding this organization: although 3 × 8 shift operation is common in industry, it is the most challenging one because it leaves you no time to take care of your equipment or to catch up in case of backlog. Equipment problems increase backlog and backlog reduces time to maintain equipment, thereby increasing equipment breakdown. To avoid getting into this death spiral, people who use such organizations tend to hide some overcapacity or inventory to deal with unexpected disruptions. Well, this is another issue; let's leave it for another day ... Let's refocus on our simulation.

From this moment, everything was almost automatic, as Daniel had done it countless times. He introduced the next point with a question, "On which tasks can Standardized Work forms be written?" As he was getting little answer from the attendees, he decided to go ahead.

---

[*] The pitch of the simulation is given in the first book of the series: Implementing Standardized Work: Measuring Operators' Performance. Please refer to this book.

> - Customer demand:
>   – 17,280 T-shirts per day
> - Operating conditions:
>   – 3 shifts of 8 hours per day
>   – 4 workstations
>   – 1 operator per workstation with relief for breaks
>   → Production time per day at each workstation:
>   = 3 × 8 × 3,600 seconds
>   = 86,400 seconds

FIGURE 4.1
Key data regarding T-shirt packing.

Standardized Work is human centered. Generally said, it can be applied wherever there is a human being. Now, for this application to make sense, it needs to be applied to an operation that will be done more than once.[*] It means that there is a need for some repetitiveness. Now, the good news is that in virtually all human operations you can find, extract, or build some repetitiveness. "Extract some repetitiveness" means that the activity can be reorganized around some repetitive tasks. Cyclic[†] operations have the highest level of repetitiveness because everything is repeated every cycle. They are the easiest tasks to be addressed when it comes to Standardized Work. Cyclic operations are very common in manufacturing. Because they are the easiest ones to address, we recommend you start with them before moving to other, so-called periodical auxiliary operations like changes, start-up, shutdown, or cleaning. The tasks I just listed do not vary a lot. They have low repetitiveness mainly because they are less frequent. I have said previously that you should start with cyclic or highly repetitive tasks because they are the easiest ones. There is another reason, which is business oriented. In effect, because those tasks are performed multitudes of times, they are ideal opportunities to achieve economies of scale. The few seconds you will save on one task will accumulate into a myriad of seconds, which translates into many dollars or Euros saved. To be concrete, let us take the example of a screwing operation and do some quick math to illustrate what I am talking about (Figure 4.2). I have noticed that a very small number of operators perform this task in 4 seconds while some others, actually almost all of them, take 5 seconds or more. I would infer that there is an opportunity

---

[*] Obviously standards can be written for nonrepetitive tasks—e.g., activities that pose high safety risk like in a nuclear plant. These do not belong to Standardized Work and therefore are out of the scope of this book.

[†] Cyclic tasks are performed every time a part is processed. Other tasks may be repetitive but not cyclic, which means that they happen after a certain number of cycles. The term "periodical" is commonly used to characterize such tasks.

Volume Effect on 1-Second Saving on a Single Task	
Saving per screwing task:	1 second
Average # of tasks per piece:	25
Average daily production per plant:	1,000
# of plants:	20
Total yearly gain (seconds):	150,000,000*
Total yearly savings (in $):	$1,458,333**

* Working days per year: 300
** Labor rate: $35 per hour

FIGURE 4.2
Economies of scale can transform a single second into huge savings in the bottom line.

> to save at least 1 second per screwing task for the sake of simplicity, let's say it is exactly 1 second per screwing operation. Now in each product we make, there are in average 25 screwing operations. On average, we produce 1000 parts per day. We have 20 plants in the region that operate on average 300 days per year. With our average labor rate, we may save up to $1.5 million per year just by saving 1 second on a screwing operation. Those very repetitive tasks are where we make our bread and butter.

Daniel paused a few seconds to let the trainees digest all those numbers. Then he proceeded:

> Now, some activities may not exhibit cyclic operations. However, in these situations Standardized Work will also have a big impact. In effect, as I mentioned previously, in the apparent complexity of those activities, repetitive patterns or tasks can be identified and be subject to Standardized Work. Let us take the example of a nurse. This is a highly trained person whose daily operation may look complex. However, if you look closely at what he or she is doing, you will identify modules of tasks that are repetitive. Examples include: washing hands, taking temperatures, taking blood pressure, taking pulse, giving injections, bathing patients, giving enemas, monitoring catheters, and a lot of other procedures. All those modules of tasks can be standardized as well. Standardized Work applied here will bring benefits in terms of safety ... I mean think about the impact of improving hand washing on nosocomial diseases. Other benefits will include improvement of quality and cost of service. Again, to summarize, the helicopter-view of some jobs may give an impression of complexity and randomness but when one looks at them closely one is able to identify repetitive modules. Moreover, in some cases, it is even possible to reorganize the work to set repetitive patterns.

28 • *Implementing Standardized Work*

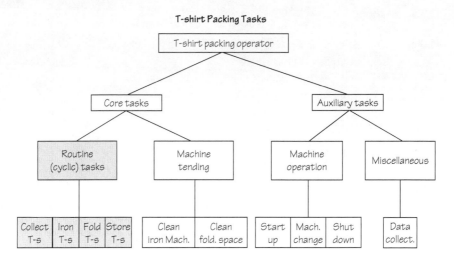

FIGURE 4.3
Breaking down T-shirt packing operation.

"Now, let us get back the task at hand: the T-shirt packing. In order for us to select the subject of Standardized Work, we must first describe all activities related to T-shirt packing. There are basically two kinds of activities: core tasks and auxiliary tasks," Daniel said while drawing (Figure 4.3).

> Let us examine core tasks. What do we have there? There are core routine tasks, which are picking, ironing, folding, and storing a T-shirt. Besides those routine activities, the operator has to do what I would call some housekeeping (clean or tidy up) from time to time on the press machine, which irons the T-shirts, and in the folding area. Now, what do we have on the auxiliary tasks side? We have, in one group, machine periodical operation tasks. Mainly, changes, start-up, and shutdown. In the other group, there is a miscellaneous task: data collection. That's all you need to do for T-shirt packing.

Daniel stopped for a few seconds and looked around to make sure that nobody was falling behind, and then continued:

> Please note that at this level, we focus on the T-shirt packing as a whole. It does not matter if the same operator or several people conduct all tasks. Now, based on our previous discussion, here is my question: How would you rank all those activities in order of priority when it comes to Standardized Work? In other words, where do you apply Standardized Work first?

On Which Tasks Can Standardized Work Forms Be Written? • 29

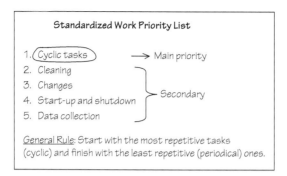

FIGURE 4.4
Order for deployment of Standardized Work.

Steve raised his hand and suggested to start with the cyclic task first and then housekeeping, changes, start-up and shutdown, and data collection (Figure 4.4). "Correct! Cyclic tasks first, then periodical ones." Daniel replied.

> I would just like to underscore that no noncyclic activity should be taken care of before all cyclic tasks of the plant are standardized. This training will only focus on cyclic tasks. I am sure that after you have taken care of all cyclic tasks, you will not need my help when you move to periodical activities. To sum up, the general rule is that you should start with the most repetitive tasks and end with the least repetitive ones. I think I have said everything I needed to say about where Standardized Work should be applied. If you have no questions regarding what we just discussed, then it is time to move to the data collection that will be used to write Standardized Work forms. We will walk through the process step by step.

Standardized Work is human centered. Therefore, generally said, it can be applied wherever there is a human being. However, it should focus on repetitive tasks. It should start first with cyclic activities. Noncyclic activity should not be addressed until all cyclic tasks of the plant are standardized. To sum up, the general rule is that you should start with the most repetitive tasks and end with the least repetitive ones.

5
Collecting Data

If there is anything Daniel had learned over years of giving training, it was the power of a step-by-step approach. He always told his young colleagues, "Break it down into small pieces, go for the step-by-step approach and people will digest it better. Moreover, it gives a lot of structure to the teaching." Training data collection was no exception to this rule. To ease the conveyance of the knowledge, Daniel crafted a step-by-step approach that he was about to unveil to the trainees.

TASK ONE: GROUP FORMING

Daniel described the first task:

> Form groups of four people. The general rule is that each group should include people with different backgrounds. When the work is done on the shop floor, please make sure that at least one operator is included. This is the most effective way to carry your message to the shop floor and spread success.

One minute later, the five groups were formed. Daniel wrote down the names on the flipchart, and then presented the layout with the position of each team (Figure 5.1). The room that was used for training was suddenly transformed into a T-shirt packaging factory with four workstations, one for each group.

32 • *Implementing Standardized Work*

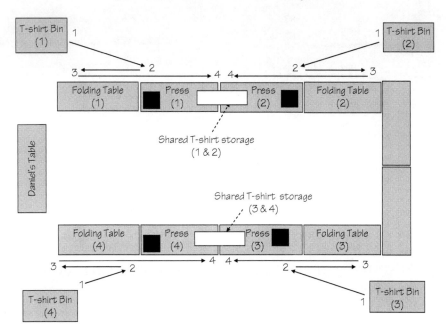

FIGURE 5.1
Layout showing the position of the four teams.

TASK TWO: TEAM ORGANIZATION

Daniel now moved to the description of the second task:

> Each team should pick an "operator." This person will be the operator during the simulation. You also need to designate a stopwatch keeper and a time collector. The stopwatch keeper will read the times loud enough for the time collector to be able to write them down on the sheet of paper I just handed out to you (Figure 5.2). The fourth person of the group, the press-handler, will be in charge of the press. Why do we need a person to be in charge of the press? The press, which is supposed to be an automated machine, is in reality just a piece of folded cardboard. Therefore, we need somebody to make it operate like an automatic machine. Every time the operator closes the press and pushes a factice button, the person in charge of the press will start the stopwatch and open the press 10 seconds later to emulate an automated operation (Figure 5.3).

Collecting Data • 33

FIGURE 5.2
Time collection sheet.

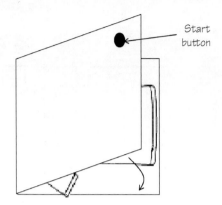

Press Made of a Folded Cardboard
1/ The operator loads the press with a t-shirt
2/ The operator closes the press and pushes the start button
3/ After 10 seconds the press-handler opens the press

FIGURE 5.3
Description of the operation of the ironing press.

A few minutes later, all teams had picked an operator, a stopwatch keeper, a time collector, and a press-handler. The whole group was now ready for the third task.

TASK THREE: FACILITATOR GIVES JOB INSTRUCTIONS FOR OPERATORS

Daniel proceeded with the description of the third task. "You have four raw material bins filled with T-shirts. They are positioned at the four corners of the room (Figure 5.1). You should pick a T-shirt in the closest raw material bin." Daniel started the demo:

> Once you pick the T-shirt, you should inspect it by checking the quality of the neckband, and the bond on the sleeve hem and the garment hem (Figure 5.4). Then iron the T-shirt in the press. This automated machine processes in 10 seconds. Once ironing is finished, fold the T-shirt and lay it in the storage area. When you fold the T-shirt, please make sure that the logo is visible as in the sample in my hands (Figure 5.4). Now I will perform two more runs; please observe what I am doing and ask questions when you need to.

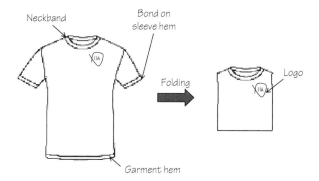

Quality Imperatives:
1/ Check the good quality of Neckband, Bond on sleeve hem and Garment hem
2/ After folding, the logo on the T-shirt should be visible

FIGURE 5.4
Quality checks to be performed on T-shirts.

TASK FOUR: TEAMS LIST THE DETAILED STEPS ON FLIPCHARTS

After performing all runs, Daniel went to the board and announced the next step, "Each group please get around your flipchart. You guys have 10 minutes to come up with a detailed description of all the steps needed to perform the T-shirt packing."

Daniel then went around from team to team to discuss their questions. Two teams were already making good progress, while the other two were still struggling to find their way. One of them had a long discussion on how detailed their description of the steps should be. Daniel explained that this was part of teamwork and that they should agree on something and execute. As he put it, "It is always better to have a 60% perfect solution executed than a perfect blueprint solution not executed." Since Daniel started delivering this training within the region, he has been amazed by the power of leadership in a group. Most of the groups with a good leader will deliver all results within the given time. A group with no leader, or too many of them, will tend to be inert or dragged down into futile discussions.

Ten minutes later, Daniel stopped the groups and asked them to share their results. Two of the four groups had around 20 detailed steps, and a third group had 25. The last group had less than 15. It happened that the group with 15 had missed some of the details of the operation. Daniel

#	Steps
1	Pick up T-shirt
2	Inspect
3	Walk to the ironing press
4	Open ironing press
5	Place the T-shirt
6	Close ironing press
7	Start ironing press
8	Wait on press
9	Open ironing press
10	Remove the T-shirt
11	Close ironing press
12	Walk to folding station
13	Fold T-shirt
14	Pick up T-shirt
15	Inspect
16	Walk to store bin
17	Store T-shirt
18	Walk to T-shirt bin and restart

FIGURE 5.5
List of the detailed steps agreed upon by the participants.

specified that the situation did not have an all right or all wrong solution. "As you will see later," he continued, "there are some gray areas where you will need to agree as a group."

After reviewing the answers of each group, Daniel asked all trainees to agree on only one set of detailed steps to be used in the next step. Thomas offered to facilitate the process, moved to the flipchart, and listed the agreed-upon steps (Figure 5.5).

TASK FIVE: FACILITATOR AND ALL TEAMS DETERMINE MEASUREMENT STEPS AND MEASUREMENT POINTS

Daniel explained that the fifth task would consist of finding two kinds of items: (1) *measurement steps*, obtained by regrouping detailed steps, and (2) *measurement points*, which define the start and the end of each measurement step. Here is how he introduced the task. "Now that we have finished with the detailed steps, it is time to introduce two concepts: measurement steps and measurement points." A quick look at the trainees'

> Measurement steps:
> - These start and end with measurement points
> - Can be measured
>   - Not too small (may be difficult to measure)
> - Are "interesting" from improvement standpoint
>   - Not too long (lack of details may hide waste)
> - Tip: estimate walk time instead of measuring (1 step = 1 second)
>
> Measurement points:
> - Things you can see
> - Things you can hear
> - Things that are cyclic (hear or see easily every cycle)

FIGURE 5.6
Summary of the definition of measurement steps and measurement points.

faces gave Daniel some clues on the amount of questions they had in their minds about those two terms. "The definitions of those two terms are somewhat intertwined," he continued while walking toward the flipchart. Then he explained, while writing down the summary (Figure 5.6),

> There is some equilibrium to be found while defining a measurement step. It should not be too short; otherwise, you will find it hard to measure. This means that you should combine enough detailed steps so that you are able to measure comfortably. For instance, a measurement step around a second could be very difficult to measure. On the other hand, you do not want your measurement step to be too long either, as it might hide precisely the waste and opportunities for improvement you are looking for.

At this point, Steve interrupted Daniel, "Daniel, what do you mean by 'hiding some waste'*?"

> This is an excellent question, Steve; I will come back to this notion of waste later. For the time being, the kinds of waste we are talking about are mostly walk and wait. Ultimately, any time spent by the worker walking or waiting will have to be visible on the Standardized Work combination table. Therefore, if we bundle too many detailed steps in a measurement step, we simply will not be able to make waste visible. Now, here is another tip that can make your life easier by helping reduce the amount of work. To reduce the number of measurement steps, you can simply count walk steps instead of measuring the time. The walk time is thereafter computed by multiplying this number of steps by the time per step. We advise you take 1 second for each step.

---
* This is part of the book dedicated to Process Improvement.

38 • *Implementing Standardized Work*

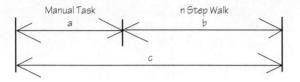

Description (what to do)	Example
Measure c	Measuring c: c = 8 seconds
Estimate b by: 1/Count the number of steps: n 2/b = n × 1 second/step	Estimating b: 1/number of steps: n = 5 steps 2/b = 5 steps × 1 second/step = 5 seconds
Deduce a by subtracting b from c: a = c − b	Deducing a by subtracting b from c: a = 8 seconds − 5 seconds = 3 seconds

FIGURE 5.7
Reducing measurement work by counting walk steps instead of measuring them.

Thomas intervened at this point: "Daniel, I am not sure I understand this point. Please, could you explain more?"

Well, let us take an example (Figure 5.7). Suppose that an operator is conducting a manual activity followed by a 5-step walk. Instead of trying to measure the manual activity and walk time, it would save time to measure the manual activity and 5-step walk as a whole. Suppose that the result is 8 seconds. Then you could infer that the manual activity takes 3 seconds. This is because 5 steps translate into 5 seconds (1 second per step). Then 8 seconds minus 5 seconds equals 3 seconds. What's more, you are saving yourself the pain of measuring two short tasks: 3 and 5 seconds. Thomas, do you have a better understanding now with this example?

Thomas nodded. Steve, the IE manager, did not look convinced. Therefore, Daniel asked him if he could do anything to clarify more. Then Steve reacted:

Daniel, it looks so approximate. I have a question regarding your tip on 1 second for a step. Based on my experience, I know that the elapse time for a step depends on a lot of factors: the height of the person, his or her pace, the weight of what he or she is holding in his or her hands, the nature of the floor. So how can you just decide that it will be 1 second for each step?

In this question, Daniel recognized the excruciating "drama" of IE people. As he often put it jokingly, "When it comes to numbers, IE guys care more about the digits after the dot than the ones before." For sure, as an IE, he could say such things without fostering furies from the IE community.

Daniel responded:

> Well, Steve, you are right. I am doing some approximation. There are two answers to your question. The short one is that 1 second is a good tradeoff between accuracy and practicality. What I mean is that taking 1 second for a step is both accurate enough and easy to manipulate. Computing round numbers is easier, especially on the shop floor where there is no calculator. That's the short answer. The long answer is that I have actually checked this number to find if there could be any link with the MTM* approach. Here is the bottom line: when you do measurement, there is really no need to go beyond the second. Because on one hand, it is time consuming. On the other hand, it has no big impact on your result. Now, you should keep in mind that the Key Point about Standardized Work is not the accuracy of the time study but the method; I mean the way people do things. Please focus on the way people are working and do not go below the second. Herein lies the stunning reality. In my experience, when you focus on time, chances are very high that you head right to a conflict with workers. If you focus on the method, you get their support. If I may paraphrase† a well-known saying at Toyota, "the right method will lead to the right time."

Daniel paused a few seconds and continued:

> Okay, we have been discussing measurement steps a lot. Now let us discuss measurement points, which, by the way, go together. Each measurement step starts with a measurement point and ends with another measurement point. These are things you can see or hear easily and instantly. Also, you want to make sure that such signals or sounds come every cycle.

Daniel paused once again for a few seconds, stared at his audience, and then proceeded:

> I hope this is clear for everyone. If not, please ask questions. Now, let us move to the practice. You all have 10 minutes to define your measurement steps by regrouping the detailed steps we agreed upon previously; also

---

* Method Time Measurement (MTM) is a time estimate method based on a standard estimate of predefined short motions.
† "The right process will lead to good results."

40 • *Implementing Standardized Work*

> *Remember that the Key Point about Standardized Work is not the accuracy of the time study but the method—the way workers do things. Please focus on finding the best way, and you will get the right time. In other words, "the right method will lead to the right time."*

define the corresponding measurement points. As always, I will be moving from team to team to answer any questions you may have.

While Daniel was walking around, he unsurprisingly noticed some differences between the four teams. One of the teams had proposed a solution with four measurement steps while the other three teams had five measurement steps. Daniel anticipated that this discrepancy would be an interesting point to share during the review of the teams' results.

Daniel focused review of the teams' results on the number of measurement steps and measurement points. First, he had each team present its results. Then, he raised the point about the different number of measurement steps (Figure 5.8) and asked the participants which of the two approaches was the right one. One participant replied:

> I believe that both are okay. Actually, the only difference between the four-measurement step solution and the five-measurement step solution is that the last step of first solution bundles two walks: "Walk to store T-shirt" and "Walk back to T-shirt bin." In the five-step solution, "walk back to T-shirt bin," the detailed step number 18, makes up the fifth step alone.

Daniel confirmed this analysis. He also underscored that the number of steps would not have so much impact because the result would be the same. For the sake of clarity, and to easily review the work of the four teams, Daniel proposed that all participants keep the four-step solution, which made it easier to measure, and he proceeded with the simulation.

Daniel went on to conclude the review with the following remark:

> While touring from team to team earlier, I noticed that some of you had more measurement points than measurement steps. Please remember that measurement points will always be both the end of one measurement step and the beginning of the next measurement step. That means you should have exactly the same number of measurement steps as measurement points. For instance, in the case of the four-step solution, the arrows on this flipchart show that you do not need more than four measurement points

Result for Team 1 (4 Steps)

Start	Measurement Steps (4 Steps)	Stop	Corresponding Detailed Steps #
Touch T-shirt	Pick up T-shirt, inspect, walk	Touch press (opening)	1, 2, 3
Touch press (opening)	Load and start press, wait, and upload press	Release press (closing)	4, 6, 7, 8, 9, 10, 11
Release press (closing)	Walk, fold, pick up, and inspect T-shirt	Pick up T-shirt	12, 13, 14, 15
Pick up T-shirt	Walk, store T-shirt, and walk back to T-shirt bin	Touch next T-shirt	16, 17, 18

Result for Teams 2, 3 and 4 (5 Steps)

Start	Measurement Steps (5 Steps)	Stop	Corresponding Detailed Steps #
Touch T-shirt	Pick up T-shirt, inspect, walk	Touch press (opening)	1, 2, 3
Touch press (opening)	Load and start press, wait, and upload press	Release press (closing)	4, 6, 7, 8, 9, 10, 11
Release press (closing)	Walk, fold, pick up, and inspect T-shirt	Pick up T-shirt	12, 13, 14, 15
Pick up T-shirt	Walk, store T-shirt	Release T-shirt	16, 17
Release T-shirt	Walk back to T-shirt bin	Touch next T-shirt	18

FIGURE 5.8
The two solutions presented for measurement steps and measurement points.

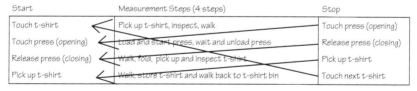

Each measurement step is the end of a measurement step and the beginning of the next measurement step.

FIGURE 5.9
There should be the same number of measurement steps and measurement points.

(Figure 5.9). One last point before we move to the sixth task. We have all agreed that we will use the four-step solution. I want to make sure that you will take the same measurement points so that we can share and compare your results later on. As you all know, "showing is always better than telling." Therefore, I suggest that the operator for team 1 perform a dry run of a full cycle and say "top" every time there is a measurement point.

TASK SIX: EACH TEAM TO PERFORM 20 REPS

Daniel began:

> Okay, you are now ready for time collection. Everyone plays his or her role. Take the Time Collection sheet I gave you previously (Figure 5.2). The first thing you should do is write down each measurement step in the dedicated cell, one measurement step per row. For this to be practical, you should name them.

Then Daniel proceeded cheerfully, "Let us agree quickly on the naming." Sarah, the HR manager, offered to facilitate the exercise, which led to quick agreement on the name of steps (Figure 5.10).

After this quick digression, Daniel went back to the time collection sheet:

> Okay, let's refocus on the time collection sheet. Now, if you move to the next column you will see that each row has two sub-rows. In the lower one, you should write down the time you read on the stopwatch. There is one important piece of information to remember: start your stopwatch on the first T-shirt and let it run continuously until the end of the 20 reps. This means that the time you read will be the cumulative time. This is why you need the upper row. It helps you calculate the actual time of each step. To fully emulate the real conditions of work on the shop floor, please do not stop your stopwatch, whatever the reason. If you encounter any disruption, write it down on the bottom of the time collection sheet just like a footnote. Specify the cause and your best estimate of the duration. You have 10 cells reserved for that. Now, if everyone is ready, let's get started.

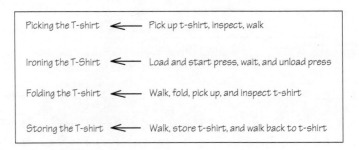

FIGURE 5.10
Naming measurement steps.

Collecting Data • 43

Time Collection Sheet

Machine:
Date: June 6, 2012
Time:
T-shirt Packing Simulation
Realized By: Team 1
Station:

Process Cycle →

#	Operation Element		1	2	3	4	5	6	7	8	9	10	11	12	13	14	15	16	17	18	19	20	Mode	Min	Max	Average
1	Packing T-shirt	Measured time:	4"	6"	7"	5"	8"	7"	7"	6"	8"	12"	6"	7"	5"	7"	10"	7"	6"	7"	8"	8"	7	4	12	7,05
2	Ironing T-shirt	Measured time:	4"	4'4"	11'28"	2'7"	2'47"	3'27"	4'7"	4'46"	5'24"	6'47"	7'25"	8'9"	8'45"	9'23"	10'9"	10'59"	11'35"	12'11"	12'50"	13'35"	11	8	14	11,4
		Measured Time:	14"	9"	11"	8"	12"	12"	12"	11"	11"	11"	13"	11"	12"	11"	11"	12"	11"	11"	14"	10"				
3	Folding T-shirt	Measured time:	18"	5'3"	1'39"	2'15"	2'58"	3'39"	4'19"	4'57"	5'35"	6'58"	7'36"	8'20"	8'57"	9'34"	10'20"	11'11"	11'46"	12'22"	13'4"	13'45"				
		Measured Time:	12"	20"b	15"	16"	13"	14"	14"	11"	13"	13"	15"	12"	11"	17"	20"c	10"	11"	12"	14"	11"	11	10	20	13,7
4	Storing T-shirt	Measured time:	30"	1'13"	1'54"	2'31"	3'12"	3'53"	4'33"	5'8"	5'48"	7'11"	7'53"	8'32"	9'8"	9'51"	10'40"	11'21"	11'57"	12'34"	13'18"	13'56"				
		Measured Time:	8"	8"	8"	8"	8"	7"	7"	8"	47"a	8"	9"	8"	8"	8"	12"	8"	7"	8"	9"	10"	8	7	47	10,2
		Measured Time:	38"	1'21"	2'2"	2'39"	3'20"	4'0"	4'40"	5'16"	6'35"	7'19"	8'2"	8'40"	9'16"	9'59"	10'52"	11'29"	12'4"	12'42"	13'27"	14'6"				
5		Measured Time:																								
		Measured Time:																								
6		Measured Time:																								
		Measured Time:																								
7		Measured Time:																								
		Measured Time:																								
8		Measured Time:																								
		Measured Time:																								
9		Measured Time:																								
		Measured Time:																								
10		Measured Time:																								
	Total Process Cycle Time																									
	Special Cause Time																									

Comments: Note reason for special cause variation, material change, code change

1	a material change, lasted 39 seconds	6
2	b. press breakdown (simulated by trainer), lasted 10 seconds	7
3	c. press breakdown (simulated by trainer), lasted 9 seconds	8
4		9
5		10

FIGURE 5.11
Time collection sheet from Team 1.

44 • Implementing Standardized Work

	Team 1	Team 2	Team 3	Team 4
Picking the T-shirt	7	21	5	21
Ironing the T-shirt	11	11	11	12
Folding the T-shirt	11	24	19	28
Storing the T-shirt	8	5	7	11
Target Cycle Time	37	61	42	67

FIGURE 5.12
Modes of the four measurement steps of the four teams.

Twenty minutes later all teams had completed their 20 reps. Daniel asked them to fill out the three last columns of the time collection sheet. He explained, "For each measurement step, please compute the mode, the minimum value, the maximum value, and the average." After each team completed the task, Daniel called all to order for review of the results (Figure 5.11). He asked the four teams to give him their modes. He wrote them down on a poster (Figure 5.12). Once again, he took this opportunity to insist that the mode was the best measurement of manual time. As he stated,

> Workers tend to easily accept the mode versus other values. The minimum repeatable value* that some Lean practitioners adore tends to antagonize workers who view it as unfair. It is not a winning approach. Besides creating a focalization on the time versus the method, you run the risk of losing workers' early buy-in to the Standardized Work deployment process.

At this point, Daniel was expecting Steve, the IE manager, to intervene. This was one of the points on which he had expressed some frustration about the Excellence System peoples' approach: "You can't just go out there on the shop floor with a stopwatch, measure a few cycles, take the minimum repeatable, and decide that this should be workers' standard time." He believed that time measurement could only be done by knowledgeable people, and those knowledgeable people were IE people, end of story. No surprise that Steve did not miss the opportunity, and, as Daniel expected, he promptly raised his hand and said, "Daniel, as you know 'a picture is worth a thousand words'; could you grant me two minutes to explain very quickly how we measure standard time?"

---

* This is the smallest of the values that appear at least twice in a list.

| Observed Time (sorted out and averaged) | Multiply By Performance or pace rating (80% to 120%) | Add Allowance For Personal needs (≈5%) | Add Allowance For Fatigue (≈4%) | Add Allowance For Unavoidable stops (≈1%) |

Standard Time (brace spanning Multiply By through Add Allowance For Unavoidable stops)

FIGURE 5.13
Standard time breakdown, according to the IE manager's definition.

"Yes, sure. Please come to the flipchart and explain," Daniel said. Steve started his drawing while talking:

Well, to determine a standard time, the first thing we do is observe a few operators. Then we remove outliers from our list of data. This is what I call sorting out. For example, if because of a breakdown, the operator spends more time than normally needed to complete his task, I will remove this cycle from my sample of values. The next step is to average the remaining values. This average value will be corrected by a factor called pace rating or performance rating (Figure 5.13). What is the pace rating? Well, every time we observe an operator and collect the time we always note an estimate of the pace of the operator. We rate 100% an operator who works at the normal pace. I have to say that for several reasons operators do not always work at the normal pace. For example, when observed, many operators with some slack time tend to slow down their pace to show that they have no flexibility. Therefore, the collected time will be multiplied by a factor below 100%. From this point, we will add different allowances. First, a 5% allowance is granted to the operator for personal needs. We also add an average of 4% allowance to allow the worker to recover from physical fatigue resulting from the conduct of his work under some specific work conditions. Of course, this is an average. The actual allowance will depend on the nature of the task and the work conditions itself. For instance, it could reach up to 10% for a worker carrying a heavy part with some ergonomic issues. Conversely, it could go down to 1% if the operator is carrying very light parts or no parts and works in the best ergonomic conditions. Finally, to complete the whole picture we add a third allowance whose value is around 1% to take into account all unavoidable stops imposed on the operator. As you may see, this is not something that could be done in a so-called "quick-and-dirty" way. That's it, Daniel.

Daniel thanked Steve while he was heading back to his seat and proceeded:

For sure, Steve, the standard time as you just explained to us is probably fairer than any other method we have discussed before. However, I would like to remind everyone that all those allowances you talked about convey the elements of agreement between employers and trade unions regarding job requirements and wages. For this same reason, its computation, as you just showed us, may depend on several factors, which ultimately become inaccessible to non-IE people. This may create a distraction from our objective. Therefore, we cannot use this standard time in workshops without running the risk of deviating from the agreed objective: "Focus on the method not the time." Always keep in mind that we want to end up with the best method. I mean the best method in terms of safety, ergonomics, quality, productivity, and cost. Our rationale being that if we apply the best method, the right time will follow.

Daniel paused a few seconds and continued:

Please do not get me wrong. I am not dismissing the need for standard time as detailed by Steve. The point I am making is that while we need this to embody agreements with trade unions, it is simply not suitable to use for Standardized Work training and workshops.

The time was right for another 5-minute break. Before Daniel announced the break time, Thomas made a sign to signify that he wanted to make a comment. Daniel reacted, "Before we take a short break I give the floor to Thomas, who would like to add something. Please, Thomas, go ahead."

Thomas moved to the flipchart and proposed to summarize the main learning:

Daniel, apart from the method for data collection, here is my takeaway from what you taught us. The one thing that popped out was that we really need to focus on the method versus the time. I understood that if we focus on the time we would get conflict and random results whereas by focusing on the method, we would get teamwork and the right results. I found this powerful (Figure 5.14).

"Thank you, Thomas, for illustrating your conclusion," Daniel said.

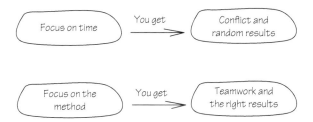

FIGURE 5.14
The right approach to Standardized Work should focus on the method, not on the time.

6

Introducing the Four Standardized Work Documents

Before getting into writing Standardized Work forms, Daniel underscored once again the need for minimal stability before the implementation of the Standardized Work:

> Again, you do not need to have 100% stability before you start writing your Standardized Work sheets. I am sure you still remember the spiral I presented to you yesterday.[*] As discussed yesterday, you need to do some "cleaning" in order to achieve minimal stability, and then jumpstart the implementation of Standardized Work. Give me a few seconds to clarify a point here. Writing Standardized Work documents simply amounts to taking a picture. In principle, you can do it any time, even before addressing stability issues. It will simply be tougher, as instability will prevent you from taking a clear picture. It is just like trying to draw a picture of a moving model. Even the best artist can fumble. However, do not go farther than the writing if your process is still unstable.

After this clarification, Daniel went back to the course of the training. "Stable process means that your machines should stop as rarely as possible." Daniel felt it was the right time to interact with the trainees on this matter. Therefore, he shifted to question mode. "What are the possible reasons for which your machine will stop?" Steve raised his hand and answered, "I think most of the reasons, at least the most important ones, belong to what we call OEE[†] losses. I mean downtimes, quality issues, and performance issues (micro stoppages, reduced speed) … That's all I can see."

---

[*] Subject explained in the first book of the The One-Day Expert series dedicated to Standardized Work: Measuring Operators' Performance.
[†] OEE stands for Overall Equipment Effectiveness.

49

"Thanks, Steve. What else? Any other ideas?" Daniel asked. He then directed himself toward the production manager and their group leaders. "Guys, can you tell me which kinds of disruptions stop your machines on the shop floor?" One of the participants, who was an assembly area group leader reacted, "Besides the causes Steve just gave, I would add people-related reasons like absenteeism, sickness, and change in the method ..." Daniel interrupted him:

> Thank you for your answers. I should have said that I meant non-human-related reasons. Okay, anyway I will keep the change in the method and go back to Steve's list. If we put this together and summarize, I would say that there are two main pre-conditions for the deployment of Standardized Work. The first set is related to the stability of the process and more especially disruption. This is related to what I called yesterday stochastic special cause variation.[*] I mean breakdown (failure of equipment, random lack of material, or random blocking of the machine for lack of storage space), and quality issues. The second condition includes the method used in the work. It must be repetitive, at least by design. I used the term "by design" because no work can actually be performed in a repetitive way if it faces frequent disruptions. Therefore, at this stage, we do not judge the work as observed, but the work as designed. We make abstractions of all disruptions.

> The next thing I would like to discuss with you is the list of the Standardized Work documents. I have already presented two of them to you during the warm-up quizzes: the Standardized Work chart and the Standardized Work combination table. In most books and teaching, people would add to these two forms the Process Capacity Sheet. My colleagues and I in the region have decided to replace it with a capacity analysis bar that some Japanese companies call Yamazumihyo. I will come back to this later when we perform the writing on the T-shirt packing simulation.

> In addition to those three documents, we have also added a fourth one in the list, which is the Operator Work Instructions.[†] So if you have followed me well, this makes a total of four documents (Figure 6.1). We think it makes no sense to just copy everything in the literature or everything that Toyota does and push it. We believe that tools should be used because they solve our problems, not because they exist or because other companies are using them. Let me repeat: focus on your problem, choose the right tools, and even adapt them, if necessary, to solve it. And guess what? This is

---

[*] More details on the classification of variation are given in the first book: Implementing Standardized Work: Measuring Operators' Performance.

[†] There are various names of this document in the literature, including Operation Instruction Sheet.

Introducing the Four Standardized Work Documents • 51

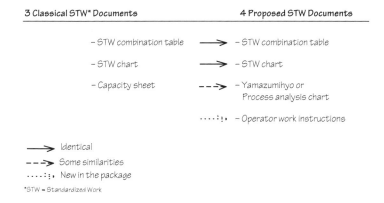

FIGURE 6.1
Modifications made in the list of Standardized Work documents.

> exactly what Toyota did in the 1950s. They copied a lot of tools and concepts from the West and adapted them to fit their business needs.

Daniel badly wanted to share his thoughts about a topic that belonged to his hobby horse. But, he was not sure that it was the right time. After a few seconds of hesitation, he decided to dive in.

> Okay, it's killing me not to talk about this. Let's take a few minutes of digression to discuss what I call "Toyota fetishism." I have to say that I am always astounded by the way people are prompt to copy and deploy a tool just because Toyota is using it. In some places, saying that something has been used by Toyota is de facto a guarantee for everyone's allegiance. I have a great respect for Toyota. I believe that they have huge knowledge in manufacturing, probably more than any other company. However, I am sick and tired of the way people just copy and paste without reflection on their own need. I have to say that it has not been easy to convey this until recently. Toyota's recent difficulties helped folks understand that Toyota, the king of operational excellence, does not always get everything right, therefore prompting more people to venture into their own ways.

Daniel had had a similar discussion with Thomas previously, and he knew that this was one additional point that they shared in common. Both were very rational people and avoided doing things without understanding the rationale. As Thomas put it:

> I believe that in order to lead people properly, good leaders should understand the "whys" and be able to explain it to their people. In case they do

not understand, they should experiment and learn from the results: this is pure and simple PDCA* approach.

After his rousing comments on so-called "Toyota fetishism," Daniel quickly refocused the group on the simulation:

If you have not followed everything regarding the list of Standardized Work documents, do not panic. This is simply the introduction. We will examine each of those documents, and learn how to write and use them later on. That said, if you have any questions thus far feel free to ask.

Daniel got no questions. Therefore, he moved to the next part of the training about how to write a Yamazumihyo chart.

> *We don't think it makes sense to just copy everything in the literature or everything that Toyota does and push it. We believe that tools should be used because they solve our problems, not because they exist or because other companies are using them. Focus on your problem, choose the right tools, and even adapt them, if necessary, to solve it.*

---

* PDCA stands for Plan, Do, Check, and Act. It is a method of resolution of problems originated from the scientific methods, formalized and advocated by Dr. W. Edwards Deming.

7
Getting Ready

Daniel told the group that the four Standardized Work forms had four key elements in common. He explained:

> In order to write these forms, we should study and prepare these common elements first. Before we go any further, I need to let you know that most of the literature on Standardized Work forms identifies three common elements. The first element is the Takt time, which we have already discussed. The second element is the Job Sequence. It describes the order in which the tasks are realized in the most effective way. I mean "effective" in terms of safety, quality, and productivity. The third element of Standardized Work is the Standardized Work in Process (SWIP). It is made of the minimum parts needed to run production as described in the Standardized Work forms. As we discussed before, we are not in the "copy-paste" business. Because Standardized Work should describe the best method of doing a job, we thought that one of the most important contributions was to convey the knowledge regarding this best method of doing a task. It happens that this knowledge, a least a big chunk of it, is conveyed through Key Points, which therefore play the role of "knowledge reservoir." As a result, besides the three elements identified by Toyota (Takt time, Job Sequence, and SWIP), we added a fourth one: Key Points. Let me underscore this. Standardized Work with no Key Points or with poor Key Points is just a paper wall. You cannot effectively train operators without good Key Points. That means no real application of the concepts and, ultimately, deployment will be a pure waste of time. Before we study the four key elements, which are Takt time, Job Sequence, Standardized Work in Process, and Key Points, I suggest you take a short break. Please be back at 10:30.

> *Standardized work with no Key Points or with poor Key Points is just a paper wall. We cannot train operators effectively without good Key Points. That means no real application and, ultimately, deployment will be a waste of time.*

THE FIRST ELEMENT OF STANDARDIZED WORK FORMS: TAKT TIME

When the group returned from break, Daniel went to the flipchart and said:

> As discussed previously, the first element of Standardized Work is the Takt time. Among all the elements we will discuss, this one is certainly the most important. I am sure that some of you know how to compute the Takt time. Okay, who in this room knows how to compute the Takt time?

Only a few people raised their hands. Among them were Thomas, Steve, Eric, the Excellence System Manager, and another participant. As Daniel anticipated, only a few people knew how to compute such an important element of Lean: four people in a group comprised of the highest ranked managers. Even though Daniel had given a definition of the Takt time earlier, he knew that that was not enough; therefore, he thought another explanation followed with an application would be highly beneficial to the audience. As he always does at this stage of the training, Daniel asked one participant to explain the Takt time computation to the other colleagues. This person was Eric, the Excellence System Manager. Since it was a key element of the Lean implementation, the local Lean authority was the right person to take over the duty.

Daniel prompted, "Please, Eric, could you re-explain to the rest of the room what the Takt time is, and most importantly, how to compute it?"

Eric replied, "The Takt time is the voice of the customer. It tells us if we are satisfying or can satisfy the customer demand. This happens when the operator can complete his cyclic work below the Takt time. It is obtained by dividing the production time by the customer demand."

Daniel continued,

> Thanks, Eric. I completely agree with what you said. I would add that satisfying customer demand in a broader way means, "delivering what is

needed, when it is needed, in the quantity requested." Also, I need to point out that when customer demand is satisfied, it is not the end of the story. Takt time helps find out whether this is achieved in a waste-free way. In other words, it helps determine whether operators are well balanced or not. More prosaically, from the practical standpoint, if you need to review the Standardized Work, as we will do tomorrow, Takt time helps you know which amount of the work can be given to an operator without hampering customer requirement. For instance, you will give more tasks to an operator when the Takt time moves from 45 seconds to 60 seconds. If you don't, you will, for sure, satisfy the customer demand at this workstation but with 25% of labor capacity wasted.

Eric responded, "True. I forgot the application of the Takt time to operator balancing in the Standardized Work definition."
Daniel continued:

Well, let us go back to the Takt time computation. As you just presented it, the Takt time formula is simple. It is the production time divided by the customer demand. The customer demand is clear: 17,280 T-shirts per day. Remember, I gave you this information earlier today. The production time is not as simple. This is where troubles start most of the time. Do you have any idea of how to get it?

Eric had a suggestion:

Well, Daniel, here is what I propose. I would take the full-day production time that we computed previously (Figure 4.1), which was, I think, 86,400 seconds. This is because we have break reliefs. If not, we would have subtracted breaks, unscheduled time, and other planned stops.

Eric, who was a little hesitant, fixed his wandering eyes toward Daniel's in a quest for confirmation.
Daniel replied,

That's correct. You have to subtract all breaks and planned stops, which are not covered. Now be careful, you should not remove any provision for OEE losses. I mean losses due to availability, which includes breakdown and changes; losses due to efficiency, which includes micro stoppages and speed loss; and losses due to quality issues. Eric, please go ahead.

Daniel talked while displaying the chart Eric mentioned so that everyone in the room could see the numbers (Figure 4.1).

Eric:

You also gave us the customer demand, which is 17,280 T-shirts per day. Based on those two numbers, we can infer that our Takt time should be 86,400 seconds divided by 17,280. I can't do this mentally; can someone do the computation for me?

Daniel:

Not necessary. That's 5 seconds. Don't be impressed. This is not the first time I have done this; I know the answer. Well, so far we can say that our customer is expecting a T-shirt to come out of our plant every 5 seconds. Is this clear for all of you? Well, it looks like 5 seconds is the Takt time at the plant level. This tells you nothing about what your customer voice at each machine level is. Got any idea, Eric?

Eric:

Well, we have four workstations that should contribute equally. Therefore, the Takt time at the machine level, if I may use this phrase, should be 5 seconds times 4, which gives 20 seconds. Correct?

Daniel:

I understand you are cautious when using the phrase "Takt time at machine level." I have had countless arguments on this point in previous trainings. Some people believed that the term Takt time could only be used for the 5-second measure. I will suggest we save our time here by avoiding getting into such sterile discussions. People spend too much time in scholarly argument on the peripheral parts of Lean concepts. My general position on this is that as long as the concept is clear, and all stakeholders know what they are talking about, the term used is secondary. As I said previously, focus on solving your own problems, own your tools, and make whatever modification or renaming you desire. This allows you to save time to focus on the most important thing: the application of the concept itself. If nobody minds here, I suggest we use the phrase "Takt time at machine level" and, for the sake of simplicity, I will even suggest that we call it Takt time.

A gaze at the room showed that all participants agreed. At this moment Thomas intervened, "Daniel, what if we summarize a few of the things we just discussed?" Daniel nodded. Then Thomas quipped, "If you do not mind, I would like to volunteer at the flipchart."

> **1st Element of STW: Takt Time**
>
> - Is the voice of the customer
> - Helps answer the questions:
>   – Are we satisfying or can we satisfy the customer demand?
>   – If so, are we achieving this with a waste-free production?
> - Calculation
>
> $$\text{Takt time} = \frac{\text{Production time per day (seconds per day)}}{\text{Customer demand (parts per day)}}$$
>
> Where
> - Customer demand is provided by the customer
> - Production time is the time during which production is made
>   – Remove breaks, unscheduled time, and other planned stops, if any
>   – Do not remove provision for OEE losses—i.e., losses due to availability (breakdowns, changes), efficiency (micro stoppages, speed loss), or quality

FIGURE 7.1
Computing the Takt time.

Thomas jumped to the flipchart and started writing while explaining (Figure 7.1):

> Here is what I understood from the broad definition of the Takt time. Now on our practical case, we divided the production time, which is 86,400 seconds per day, by the customer demand, which is 27,280 T-shirts per day. This gave us the Takt time at the plant level. We then multiplied the result by 4 to obtain 20 seconds, which is the Takt time at each of the four machines. Finally, we renamed this "Takt time at machine level" into Takt time.

Daniel thanked Thomas for the summary and completed, "I hope you have noticed that we are using seconds as time units here. This is for the reason I explained early today. Seconds makes things more accessible to people."

To conclude the segment on the Takt time, Daniel insisted:

> Again, this is the most important element of Standardized Work because it sets the time allowed for the worker to perform his or her operation. Hence, it has a big impact on the definition of the operator task. For instance, if the Takt time moves from 60 seconds to 90 seconds, each operator will have more time to accomplish his or her activity. That means a new way of working with different job sequences and a potentially different Standardized Work in Process. This means virtually everything else, except for Key Points, changes when Takt time changes.

> *People spend too much time in scholarly argument on peripheral parts of Lean concepts. As long as the concept is clear and all stakeholders know what they are talking about, the term used is secondary. Focus on solving your own problems, own your tools, and make whatever modification or renaming you desire. This allows you to save time and focus on application of the concept itself.*

THE SECOND ELEMENT OF STANDARDIZED WORK FORMS: JOB SEQUENCE

Daniel moved on to the next topic:

> The second element of Standardized Work is the Job Sequence, which describes the most effective order for workers to realize tasks. Let me write it down (Figure 7.2). Job sequence is made of *major steps*. Major steps are derived from measurement steps by suppressing everything that is not value-added. For instance, waiting and walking cannot be major steps. One more important point—major steps should be succinct and stated in the following way: action + object.

Daniel paused, breathed, and continued. "If everything is clear, I'll let you define the major steps." After a few minutes, Daniel reviewed the results with each group. He asked Sarah, the HR manager who had not been talking a lot, to facilitate and converge to common answers (Figure 7.3).

THE THIRD ELEMENT OF STANDARDIZED WORK FORMS: STANDARDIZED WORK IN PROCESS (SWIP)

"As discussed earlier, the SWIP is the minimum number of unfinished parts necessary to produce as described in the Standardized Work method." Daniel said. He continued:

2ⁿᵈ Element or Standardized Work:
Job sequence

- Job sequence defines the most effective order of tasks:
  – Safest*
  – Best in terms of quality**
  – Most productive

- Job sequence is defined by the major steps

*Most important criterion
**Second most important criterion

Home Appliance Inc.

| Title | Bladder Assy | | | By: | | Mic |

N°	Major Steps	Time				10
		Manu	Auto	Wait	Walk	
1	Apply Component 1 & validate machine	11	41	13	3	
2	Get & apply Component 4	5		10	2	
3	Apply Component 2	17				
4	Apply Component 3 & validate machine	37	50		3	
5	Get Beads	4			2	
6	Load Beads	4			2	
7	Apply barcode & inspect FG	7		22	6	

FIGURE 7.2
Job sequence defines the most effective order of tasks through major steps.

Measurement Steps		Major steps
Pick up T-shirt, inspect, <u>walk</u>	⟶	Pick up and inspect T-shirt
Load and start press, <u>wait</u>, and unload press	⟶	Load, start, and unload press
<u>Walk</u>, fold, pick up, and inspect T-shirt	⟶	Fold and inspect T-shirt
<u>Walk</u>, store T-shirt, and <u>walk back to T-shirt bin</u>	⟶	Store T-shirt

FIGURE 7.3
Going from measurement steps to major steps—1: only value added tasks and 2: succinct formulation in terms of "action + object."

The main goal is to avoid the classical tendency of a worker to accumulate a lot of unfinished parts at the workstation; the direct consequence will be a disruption of Standardized Work. This can even lead to unsafe acts or production of defective parts. We advise you to represent SWIP parts of the Standardized Work chart in stations and not in raw material or finished goods storage locations. To illustrate my words, let me draw a very quick process flow chart of the T-shirt packing operation (Figure 7.4). As you may see, for the process to be run as currently defined, we need only one part in the flow. This unfinished part is represented in the machine, which is a value-adding position.

60 • *Implementing Standardized Work*

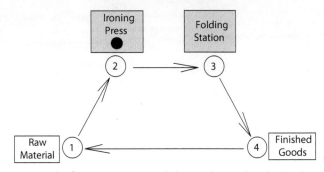

- SWIP is made of minimum parts needed to produce as described in the Standardized Work forms
- Parts should be represented in an adding value position: station or machine, which are in gray

FIGURE 7.4
Standardized Work in Process is the minimum number of unfinished parts needed.

THE FOURTH ELEMENT OF STANDARDIZED WORK FORMS: KEY POINTS

Daniel started this section with a caveat, wherein he explained again the reason for considering Key Points as an element of Standardized Work.

> Key Points are not generally mentioned as an element of Standardized Work but are actually an important item of Standardized Work. They capture the indispensable knowledge that operators need to perform their work effectively, which is actually the rationale for Standardized Work implementation. This is the reason why we have decided to make the Key Points the fourth element of Standardized Work. Key Points can be related to many items: safety, quality, productivity, cost, ergonomics, environment, you name it.

Daniel advised that the attendees focus on the following key items:

- First, safety, the single most important item that trumps the next two others.
- Second, quality, the next most important one.
- Third, productivity. Please note that productivity Key Points may be called *tips*.

At this point, a trainee raised his hand and asked Daniel if he could restate "What makes a point a Key Point and what is the method to find them." Daniel thanked the participant for his question and then responded:

> Look, this is a common question I get on Key Points. I am glad you asked it. It gives me the opportunity to clarify. As discussed earlier, you can say that an instruction is a Key Point when it is clear that if it is not respected, the goal will not be achieved. For instance, it could lead to production of a defect in the case of a quality Key Point. Now, let me take a few real life examples. The first example that comes to my mind is opening a hotel room with a swiping card. I am sure that you guys have had the experience of trying to open a hotel room door that would not open unless you swiped your card at a certain speed: not too fast, not too slow. In these conditions, getting the right speed is a Key Point in the success of your door opening operation. Some hotel receptionists used to specify this while handing out the cards. It happened a lot when this technology first began to spread. Today, it is much easier for most people to achieve the right speed. The explanation alone is a whole question unto itself and best left for another time, when we discuss training matters. The other example I can think of is cooking, especially baking. I have heard countless stories about a Key Point to be mastered in order to deliver a perfect cake. Cake cookbooks are full of tips that pretend to do that. The best story I have come across regarding baking is one Brian Joiner recounts in one of his books.[*] This story is printed on a sheet included in the pile I handed out to you this morning. The title of the story is "The Popovers." I also sent this separately last week for you to read before this training. I hope you read it. Okay, who read it? Well, I can see that almost everyone here has read the story.

Sarah, the HR manager, said she liked the story and offered to read it to the few participants who did not have the time to read. She then started:

> The Popovers, by Brian Joiner. A pastry chef's favorite confection was popovers. On a good day, he made the best popovers in the Midwest. But not every day was a good day. In desperation, he had tried all manner of things to improve his performance: He rigorously stirred the batter by hand rather than using an electric mixer. He preheated the oven to just the right temperature. He made sure the eggs were at the proper temperature, the flour of the highest quality. But no matter what he did, not every day was a good day. Then at a class he was taking, he learned about how to efficiently run experiments to find out which of the factors might affect the quality of

---

[*] Brian L. Joiner, *Fourth Generation Management: The New Business Consciousness*, McGraw-Hill, 1994.

popovers. He tested the temperature and the size of the eggs, the temperature of the batter, the quality of the flour, whenever he stirred the batter by hand or with a mixer, the temperature of the oven, and so on. He found that only one factor really made the difference: having the batter at room temperature before putting the popovers in the oven. When that factor was controlled properly, his popovers came out perfectly time after time.

Daniel thanked Sarah, and commented:

"Having the batter at room temperature before putting the popovers in the oven" is a good example of a Key Point. As Brian Joiner commented in his book, "this pastry chef identified the most critical aspect in creating perfect popovers, and at the same time, he also discovered what was *not* important. Now he didn't have to wait for the eggs to warm up when he took them out of the refrigerator ... and all those years he had wasted time stirring the batter by hand when the mixer would have worked equally well! Now he also knew that he could revert to manual mixing if the mixer ever broke down, with no effect on the quality of the popovers." The Key Point here is an important piece of information that helps you focus on what is important, and at the same time, release or save your effort on non-important points. Once again, because they are so important, we believe that they belong to key elements of Standardized Work.

After this illustration of Key Points on a real life example, it was time for application. Daniel gave the group 10 minutes to come up with the Key Points regarding the T-shirt packing process. This was a very challenging exercise for the group. A Safety Key Point related to the press emerged very quickly from all teams. However, three of the four teams stated the Key Point in a negative way. One of the teams wrote, "Do not keep your hands in the press." Daniel took this opportunity to explain that all Key Points should be stated in a positive way. He remarked:

The key is about what to do in order to be successful, and not what not to do because there are too many other ways of "what not to do" and only one way of "what to do." Please focus on how to find the best, simplest way to describe it, as well as the right drawing or picture to accompany the description. We will come back to pictures and drawing when we discuss Standardized Work forms writing.

Daniel noticed that none of the teams mentioned any "Tips," which is the word used to refer to a Key Point related to productivity coming from

1. Pick up and inspect T-shirt	Safety Key Point: none	
	Quality Key Points:	
	• Sleeve and garment hems must be flat	
	• The neck line should rest flat against the body	
	• The neck line should recover properly after being slightly stretched	
	Tip: none	
2. Load, start, and unload press	Safety Key Point:	
	• Maintain your hands out of the press while it is closing	
	Quality Key Point: none	
	Tip: none	
3. Fold and inspect T-shirt	Safety Key Point: none	
	Quality Key Points:	
	• Inspect the overall quality of the folding	
	• Check that the logo is visible on the front side of the T-shirt	
	Tip: none	
4. Store T-shirt	Safety Key Point: none	
	Quality Key Point: none	
	Tip: none	

FIGURE 7.5
Key Points agreed upon by the group for the T-shirt packing process.

ease of work or time saving. He commented that the trainee might have needed more time and more expertise about the process to identify valuable Tips to be included in the document. He then asked the group to agree on the Key Points to be carried over to the next stage of the training. This time Eric volunteered to facilitate (Figure 7.5).

Daniel also advised the group to avoid listing too many Key Points. "Too many Key Points will kill Key Points," he said, suggesting that if there are too many Key Points mentioned then they will become banal and lose the attention of the operators. Answering a participant who asked, "How are we going to know the right number of Key Points?" Daniel said:

> This is just like determining the Key Point itself. The only scientific method I can think of would be to perform Design of Experiments (DOE) as it is mentioned in Brian Joiner's story on the popovers. But it is a very time-consuming and heavy tool that could divert you from your objective of deploying Standardized Work. The next best way to align people when there is no easy rational and logical way to find a solution is to discuss around a table to reach an agreement. Once again, as I said previously, Key Points can evolve based on practice and new learning. Therefore, here is my advice—get together around the table with operations people (including

workers) and experts (safety, quality, ergonomics, and process), agree on something, and improve next. You will get good at identifying Key Points as you become experienced. Let me underscore this again—please involve people, especially operators. Operators are big suppliers of Key Points. The irony is that some of them are not even conscious of those Key Points. That's why you need to be good at interviewing them. We will come back to this point on Friday when we discuss training.*

Daniel stopped a few seconds and then concluded, "Before we wrap up this module and move to the practice of writing of Standardized Work forms, I suggest we make a summary." He went to the flipchart, and wrote down what he jestingly called the Key Points on Key Points (Figure 7.6). He then interacted with the participants to sum up the results obtained from the preparation phase (Figure 7.7). He listed the results of each of the four elements of Standardized Work: Takt time, Job Sequence, Standardized Work in Process, and Key Points.

- Main areas: safety, quality, and productivity
- Any point that is key in the success of a task (e.g., nonrespect of quality Key Point will lead to a defective part)
- No simple rational way to find them; based on expertise and knowledge and should be agreed by the group
- Can evolve over time based on knowledge increase from practice
- It is the "knowledge reservoir" because "Knowing what is not important is as valuable as knowing what is important"
- Money and time are saved by focusing on Key Points only

FIGURE 7.6
Key Points are an important element of Standardized Work.

---

* Subject of a subsequent book on training in the same series.

Summary of the Preparation for STW Writing

Element	Result	
Takt Time	20 seconds (at machine level)	
Job Sequence	1 – Pick up and inspect T-shirt 2 – Load, start, and unload press 3 – Fold and inspect T-shirt 4 – Store T-shirt	
Standardized Work in Process	1 part located in the machine	
Key Points	1 – Pick up and inspect T-shirt	<u>Safety Key Point</u>: None <u>Quality Key Points</u>: – Sleeve and garment hems must be flat – The neckline should rest flat against the body – The neckline should recover properly after being slightly stretched <u>Tip</u>: none
	2 – Load, start, and unload press	<u>Safety Key Point</u>: Maintain your hands out of the press while it is closing <u>Quality Key Point</u>: None <u>Tip</u>: none
	3 – Fold and inspect T-shirt	<u>Safety Key Point</u>: None <u>Quality Key Points</u> – Inspect the overall quality of the folding – Check that the logo is visible on the front side of the t-shirt <u>Tip</u>: none
	4 – Store T-shirt	<u>Safety Key Point</u>: none <u>Quality Key Point</u>: none <u>Tip</u>: none

FIGURE 7.7
Summarizing preparation results for each of the four elements of the Standardized Work.

8
Writing the Forms

INITIATION TO THE YAMAZUMIHYO OR PROCESS ANALYSIS CHART

Daniel wanted to start the presentation of Standardized Work documents by talking about what he called this bizarre Japanese word he had been using thus far without much explanation. He started:

> I saw some of you raising your brows when I used the Japanese word *Yamazumihyo* during the presentation of today's agenda. Maybe you were thinking "Oh another Japanese word!" Look, I know in the Lean community it is not uncommon for people to use Japanese words to create a sense of secrecy or belonging to a small inner circle of a happy few in the know. Frankly, I am not a big fan of this approach because the ultimate objective is to communicate the knowledge. So if we use Japanese words to disrupt or block communication, then that becomes counterproductive. Therefore, if you will, feel free to use indifferently the word Yamazumihyo, its short Yama, or Process Analysis Chart. By the way, does anyone know the meaning of Yamazumihyo?

Faced with no reaction, Daniel proceeded:

> I suspect the answer is No. Okay, "Yamazumihyo" is actually the concatenation of three words in Japanese: "Yama," which means Mountain; "Zumi," which means Accumulate; and "Hyo," which means Chart. Please note the word Yamazumihyo is not exclusively used to refer to Process Analysis charts. In some instances, it is used to refer to an operator balance chart as well.

68 • *Implementing Standardized Work*

Daniel got the feeling that these quick explanations and comments were well received and contributed to making the trainees feel more at ease. Visibly pleased, he playfully refocused on the core of the training:

> Now that you have another Japanese word in your vocabulary, let us get back to our business. All right, I have said previously that my colleagues and I had introduced the Yama chart to replace Process Capacity Sheets. Before we discuss the Yama, I would like to summarily present the Process Capacity Sheet and explain why we believe that the Yama tool is more suitable in a Standardized Work package than the Process Capacity Sheet. Do not get me wrong: I would like to be clear here. There are no bad or good tools. There are only well-adapted and ill-adapted tools. As I said before, there are many tools out there. You need to figure out which ones best suit your problem, pick them up, and if necessary, customize them. Then apply them to your problem. The situation that we would like to avoid is one where you find yourself with a "hammer" and end up seeing "needles" in all your problems.

After those introductory words, Daniel moved to the flipchart and started a comparison between the Process Capacity Sheet and the Yama chart.

> A Process Capacity Sheet focuses mainly on the determination of machine capacity and has no interest when the workstation is manual (Figure 8.1). Also, it makes sense when there are several machines through which parts are flowing. This is simply what we call a line. In such a situation, Process Capacity Sheets will help find the bottleneck, which sets the maximum capacity of the line. From the practical standpoint, a Process Capacity Sheet is less visual. In effect, it is a table filled with a bunch of data. The three

Process Capacity Sheet Attributes	Yama Chart Attributes	Yama vs. STW Philosophy
Focus on machine capacity	Is a human-centered approach	OK
Not effective for fully manual or very manual processes	Centered on manual activity and its interaction with machines	OK
Not visual, table filled of numbers	Very visual	OK
Sees the process flow thru a line	Sees the process flow thru an operator	OK
Limited to maximum capacity	Capacity (max, average, …) as organized and variability	OK

FIGURE 8.1
Standardized Work, Process Capacity Sheet, and Yama chart.

points listed previously are in direct contradiction to the Standardized Work spirit. On the contrary, the Yama chart seems to be more in sync with Standardized Work philosophy (Figure 8.1). This is why we decided to include it in the document package of Standardized Work.

At this moment, Eric, the Excellence System manager, asked Daniel if he could write a Process Capacity Sheet for the packing process of the simulation. Daniel responded:

> Well, this exercise will have no real interest as we only have one machine. If you remember, I told you that Process Capacity Sheet helps determine the bottleneck and the capacity. In our case, the press is the only machine; therefore, it is the bottleneck. The capacity or, to be more accurate, the maximum capacity of the press is straightforward.

Daniel took a marker and went to the flipchart to make the calculation (Figure 8.2).

> As always, I will use Team 1's data. To deliver each T-shirt, we need an auto time, which is 10 seconds. This is the press time. We need a loading and unloading time, which is 1 second based on our measurement results (Figure 5.12). Finally, we need a portion of change time. Team 1 has had 39 seconds of change time to produce 20 T-shirts, which makes 1.95 seconds per T-shirt (Figure 5.11). Remember, there are break reliefs; therefore, the total production time is 86,400 seconds. That makes a maximum capacity per workstation of 6671 T-shirts per day. The Process Capacity Sheet will look like the table I just drew (Figure 8.3). Again, please note that I did

$$\text{Max capacity} = \frac{\text{(capacity per Station)}}{\underset{= 10\,\text{sec}}{\text{Auto time}} + \underset{= 1\,\text{sec}}{\text{Man time}} + \underset{= 1.95\,\text{sec}}{\text{Change time/part}}}$$

$$= \frac{86,400}{10+1+1.95}$$

$$= 6671\,\text{T-shirts/day/workstation}$$

Note:
Auto time is the press time = 10 seconds
Man time is the loading and unloading time = 1 sec (based on measurement)
Change time is material change time per part = 39 sec every 20 parts = 1.95 sec

FIGURE 8.2
Computation of the maximum machine capacity of the T-shirt packing process.

70 • Implementing Standardized Work

Line: T-shirt Packing			Process Capacity Sheet				Date: June 6, 2012		
Product: T-shirt							By:		

Process Step Number	Process Name	Station Number	Manual Time	Auto Time	Cycle Time	Number of Pieces/Change	Change Time	Change Time per Piece	Max Daily Capacity
1	Press	1	1	10	11	20	39	1.95	6671
Only one process step	Only one process in the sequence	One workstation: the T-shirt packing station	This is the sum of all manual times needed to load and unload press	This is press time	Sum of manual and auto times	The number of parts between changes	Duration of a change	Change per part: 39/20 = 1.95	Maximum production

FIGURE 8.3
Process Capacity Sheet for the T-shirt packing based on Team 1's data.

not include other manual processes. It would have been for pure aesthetics because the Process Capacity Sheet focuses on machine capacity and not manual activities. It should include more than one machine. As I said before, the main information from the Process Capacity Sheet states that each press machine can process up to 6671 T-shirts per day, which totals to 26,684 T-shirts per day for the four workstations. The bottom line is that we have the required machine capacity to produce the 17,280 T-shirts per day requested by the customer. Once again, it does not tell us if our current process organization will allow us to satisfy the customer demand.

Daniel then concluded:

At this stage of the process, this is not so useful to us. The Yama chart will give you information on the capacity of your whole process as it is currently organized, not only the machine capacity. I mean average and maximum capacity. Moreover, it gives you information on the variability of your process. This is the reason why we call it Process Analysis Chart. Let us look at this tool a bit closer.

> - Shows capacity:
>   – Customer demand (Takt time)
>   – Current production (average cycle time)
>   – Current maximum production (standard cycle time or target cycle time)
> - Shows variability and impact:
>   – Deterministic special cause variation: Material, functional and code/size changes
>   – Stochastic special variation: breakdown and other process instabilities
>   – Range snapshot: difference between the longest cycle time (without deterministic variation) and the standard cycle time

FIGURE 8.4
Information provided by the Yamazumihyo chart.

As I said previously, several Japanese companies use the Process Analysis Chart, or the Yama chart. This bar in itself gives you a lot of information regarding the process at the workstation. It shows how you are performing versus your customer demand and very precisely how your customer's Takt time stands versus your average production capacity and your maximum production capacity. As we will see later, this maximum production capacity is calculated based on the measured standard cycle time, also called *target cycle time*. In the rest of the training, I will indifferently use standard cycle time and target cycle time to refer to the measured cycle time obtained as we did previously by adding the modes of all measurement steps (Figure 5.12). This bar also shows the contributions of the main categories of variation.[*] There are three items of interest. First, variations that are the easiest to address are displayed. If you remember what we learned yesterday,[†] I am talking about deterministic special cause variation. The bar shows the details split into two main types: material on one side and functional, code, or size changes on the other side. Second, it shows the total estimate of stochastic variation, which includes breakdown and other process instabilities. Third, it depicts the variability range. This is the difference between the longest cycle time (without deterministic variation) and the standard or target cycle time.

Daniel felt that he needed to be more concrete now or risk losing his audience. Therefore, he announced:

> Okay folks, let us now apply this to the T-shirt packing. Let's get started. I will walk you through your very first chart. Also, I ask that you use the same poster. I would like to see all your bars side by side. However, leave a

---

[*] See the first book of the series: Implementing Standardized Work: Measuring Operators' Performance for details regarding categories.
[†] See the first book of the series: Implementing Standardized Work: Measuring Operators' Performance for details regarding categories.

fifth column for the improved version of the chart. This part will be completed tomorrow during the module on Process Improvement.* Let us go step by step. Here are the six steps you should follow.

Daniel listed the following six steps (Figure 8.5 and Figure 8.6).

Step 1: The first thing I ask you is to compute the average cycle time and depict it as a dot on the bar.

Step 2: Draw the Takt time line. It will cross all four bars. In case you have forgotten, we computed the Takt time not long ago and found 20 seconds.

Step 3: Draw the bar representing the target cycle time as measured. I mean the sum of the time you obtained by summing up the modes of the four measurement steps. If needed, refer to the adequate chart to find the data (Figure 5.12).

Step 4: Compute material change per part, and stack it on the previous bar.

Step 5: Compute code or size change per part and stack it. If you remember very well, we had no code or size change. Therefore, this is zero.

Step 6: Compute the cycle time range beyond target. It is the maximum of values obtained from cycle time minus target cycle time minus changes† (code and material). It really gives the snapshot of all random variation.

Remember, Daniel continued, yesterday we saw that all kinds of changes were deterministic as they always come as no surprise. Step 4 and Step 5 show the contribution of all deterministic variation like changes. "Practical advice: please use your Time Collection Sheet (Figure 5.2) to do all the computation needed for Step 6. It will be easier for you (Figure 8.7)," Daniel recommended.

After Daniel defined tasks to be done, all teams went through the tasks and Daniel reviewed their work after each step. As usual, he used Team 1's data to illustrate all steps (Figure 8.5 and Figure 8.6).

Thomas looked at the charts and found them very expressive and full of learning. He wanted to hear more about them from Daniel. Therefore, he asked, "Daniel, could you tell us what can be learned from these charts?" Daniel despised handing out answers. Thomas knew that and was therefore not surprised when Daniel asked back his own questions, "Okay guys, could you tell me what kind of information you are getting from these charts?" Thomas restrained from answering, and let his staffers react.

---

* Addressed in a subsequent book in the same series.
† All deterministic variations that might be identified in the process should be subtracted.

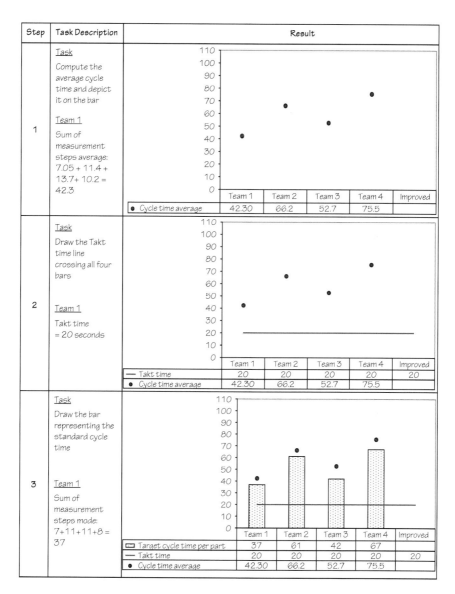

FIGURE 8.5
First three steps for building the Yama chart.

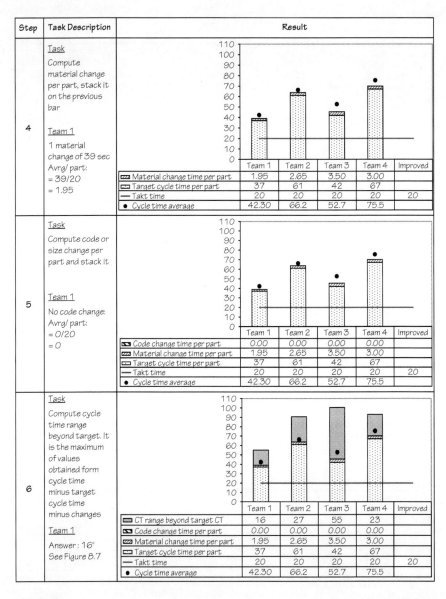

FIGURE 8.6
Last three steps for building the Yama chart.*

---

* We adopted the format used by SRI (Sumitomo Rubber Industries).

Time Collection Sheet

Machine:
Date: June 6, 2012
Time:
T-shirt Packing Simulation
Realized By: Team 1
Station:

#	Operation Element	Process Cycle	1	2	3	4	5	6	7	8	9	10	11	12	13	14	15	16	17	18	19	20	Mode	Min	Max	Average
1	Picking T-shirt	Measured time	4"	6"	7"	5"	8"	7"	7"	6"	8"	12"	6"	7"	5"	7"	10"	7"	6"	7"	8"	8"	7	4	12	7.05
		Measured Time	4"	4'4"	1'28"	2'7"	2'47"	3'27"	4'7"	4'46"	5'24"	6'47"	7'25"	8'9"	8'45"	9'23"	10'9"	10'59"	11'35"	12'11"	12'50"	13'35"				
2	Ironing T-shirt	Measured time	14"	9"	11"	8"	12"	12"	12"	11"	11"	11"	13"	11"	12"	11"	11"	12"	11"	11"	14"	10"	11	8	14	11.4
		Measured Time	18"	53"	1'39"	2'15"	2'59"	3'39"	4'19"	4'57"	5'35"	6'58"	7'38"	8'20"	8'57"	9'34"	10'20"	11'11"	11'46"	12'22"	13'4"	13'45"				
3	Folding T-shirt	Measured time	12"	20b	15"	16"	13"	14"	14"	11"	13"	11"	15"	12"	11"	17"	20c	10"	11"	12"	14"	11"	11	10	20	13.7
		Measured Time	30"	1'13"	1'54"	2'31"	3'12"	3'53"	4'33"	5'8"	5'48"	7'11"	7'53"	8'32"	9'8"	9'51"	10'40"	11'21"	11'57"	12'34"	13'18"	13'56"				
4	Storing T-shirt	Measured time	8"	8"	8"	8"	8"	7"	7"	8"	47a	8"	9"	8"	8"	8"	12"	8"	7"	8"	9"	10"	8	7	47	10.2
		Measured Time	38"	1'21"	2'2"	2'39"	3'20"	4'0"	4'40"	5'16"	6'35"	7'19"	8'2"	8'40"	9'16"	9'59"	10'52"	11'29"	12'4"	12'42"	13'27"	14'6"				
5		Measured Time																								
6	Cycle time	Measured Time	38	43	41	37	41	40	40	36	79	44	43	38	36	43	53	37	35	38	45	39				
7	Target Cycle time	Measured Time	37	37	37	37	37	37	37	37	37	37	37	37	37	37	37	37	37	37	37	37				
8	Deterministic variations (Changes) time	Measured Time									39															
9	Cycle time beyond target cycle	Measured Time	1	6	4	0	4	3	3	-1	3	7	6	1	-1	6	(16)	0	-2	1	8	2				
10		Measured Time					Cycle time range beyond target = max (cycle time − Target cycle time − Deterministic variations)																			
	Total Process Cycle Time																									
	Special Cause Time																									

Comments: Note reason for special cause variation, material change, code change

1	a. material change, lasted 39 seconds	6
2	b. press breakdown (simulated by trainer), lasted 10 seconds	7
3	c. press breakdown (simulated by trainer), lasted 9 seconds	8
4		9
5		10

FIGURE 8.7

Computing the cycle time range beyond target.

Eric, the Excellence System manager, was the first to respond:

> Well, the obvious thing I can see is that we are not meeting our customer demand. The Takt time is way below the average cycle time. The second piece of information is that there are some variations in the process. However, reducing them will not be enough to supply the customer. We will need more because the target cycle time is well above the Takt time. Based on what you have taught us, I would say that change time and equipment reliability need to be reduced to gain operator buy-in and start the process, but the method used by our operators will need to be improved to achieve the customer's demand.

Daniel was really delighted by Eric's answers. "Many thanks, Eric. I think you have drawn the main conclusions. For your information, the improvement of the method used by operators that you mentioned will be addressed tomorrow.* Who else would like to comment?" Steve, the IE manager, jumped in this time:

> I can see that there are big differences between the four teams' target cycle times. Team 1 seems to be the best of all so far. They also seem to have the lowest level of variability. When the time comes to improve our process, I guess we might have to learn a few things by observing them. Team 3 is an interesting case. They have the second best target cycle time and yet they seem to endure the highest level of variability. First, Team 3 has the highest cycle time range beyond cycle time target and the biggest gap between average cycle time and target cycle time. Above all, they have the biggest material change time per part. This really looks strange to me and I can't wait to understand why.

The questions raised by Steve were of high interest. Daniel had the explanations Steve was looking for, but it was not the right time to discuss them. He therefore thanked Steve and promised that they would find answers to his questions the next day. He also underscored that the purpose of that day's training was to learn how to write Standardized Work forms and insisted the use of those documents to improve the process and improvement actions to be inferred from them would be the subject of the next day's training.†

Thomas, who had been refraining from talking, could not stand it anymore, "Daniel, please, could you clarify a bit what you call Cycle Time

---

* This is the subject of the book of the series dedicated to process improvement.
† This is the subject of the book of the series dedicated to process improvement.

Range and its computation method?" Daniel knew exactly what was bothering Thomas. He therefore decided to get right to the point:

> Thomas, you are probably annoyed by on the use of the maximum function in the calculation (Figure 8.7). Your point being that because breakdown occurrences and duration are aleatory, a long one may happen during the observation with the consequence to "unfairly" increase the range.

Thomas acquiesced. Daniel continued:

> This is true. Alternatively, there is some probability that this situation happens. Now, the real point is that if you compare two processes, you will most probably find that the cycle time range of the most stable one is the smallest of both. The calculation, compared to other possible measurements of the variability, is simpler and yet mathematically correct. Thomas, you know better than I do that in order to survive on the shop floor, our computation needs to make sense to the people using it. For this to be possible, there are three conditions. First, it has to be simple to understand; like the notion of range to measure variability impact. Second, it needs to be easy to implement; like the usage of the maximum function. Third, and most importantly, it has to be true. Achieving all three of these criteria is never easy. However, this goal must not be relinquished. It has to be pursued relentlessly by all manufacturing support functions.

Thomas nodded while Daniel was finishing his sentence. These words resounded very well with someone who had spent a big part of his professional life trying to make things happen on the shop floor. He saw the value of this approach, and could not agree more with Daniel.

This was the end of the part dedicated to the Yama chart. Thomas did not request any summary; he noted that two of the charts they had drawn (Figure 8.5 and Figure 8.6) were not only self-sufficient, but also very good summaries of the six steps of Yama chart building.

It was getting close to noon. To avoid the crowded lunchtime, which as in most plant restaurants was at noon, Daniel suggested that the group take a lunch break. Before they left, he gave them a preview of upcoming activities:

> When you return from lunch, we will complete the three remaining Standardized Work forms: Standardized Work Combination Table, the Standardized Work Chart, and the Operator Work Instructions. Next, we will hit the shop floor, where we will apply the learning on a real workstation.

> *There are no bad or good tools. There are only well-adapted and ill-adapted tools. As I said before, there are many tools out there. You need to figure out which ones are the best suited to solve your problem, pick them up, and, if necessary, customize them, and then apply them to your problem. The situation that we would like to avoid is the one where you find yourself with a "hammer" and end up seeing "needles" in all of your problems.*

> *In order to survive on the shop floor, our computation needs to make sense to the people using it. For this to be possible, there are three conditions. First, it has to be simple to understand, like the notion of range to measure variability impact. Second, it needs to be easy to implement, like the usage of the maximum function. Third, and most importantly, it has to be true. Achieving all these three criteria is never easy. However, this goal must not be relinquished. It has to be pursued relentlessly by all manufacturing support functions.*

WRITING THE THREE REMAINING STANDARDIZED WORK FORMS

The lunch break was very short and the group soon returned to the room, where Daniel restated the next tasks:

> Let us do it step by step also. You have 30 minutes to write the remaining three Standardized Work forms: the Standardized Work Chart, the Standardized Work Combination Table, and the Operator Work Instructions. Normally, this should be no problem because you have readied all the four elements of the Standardized Work. However, I will be around, should you need my support, let me know.

As Daniel anticipated, the group worked quite smoothly. He got very few questions. The main difficulties came from the combination table (Figure 8.8). They had to determine all sorts of times: manual, auto, wait, and walk. Some teams had forgotten to count the number of steps walked by the operator when moving from one position to the next. Another difficult moment came when they needed to represent auto and wait times on

Standardized Work Combination Table

Title

By: Team 1

Takt Time:	20 seconds	Comments
Volume:	17,280 T-shirts per day	
Date:	June 6, 2012	

N°	Major Steps	Time				Time Graph (seconds)
		Manu	Auto	Wait	Walk	5 10 15 20 25 30 35 40
1	Pick up and inspect T-shirt	4			3	
2	Load, start, and unload press	1	10	10	2	
3	Fold and inspect T-shirt	9			2	
4	Store T-shirt	2			4	
	Total	16	10	10	11	

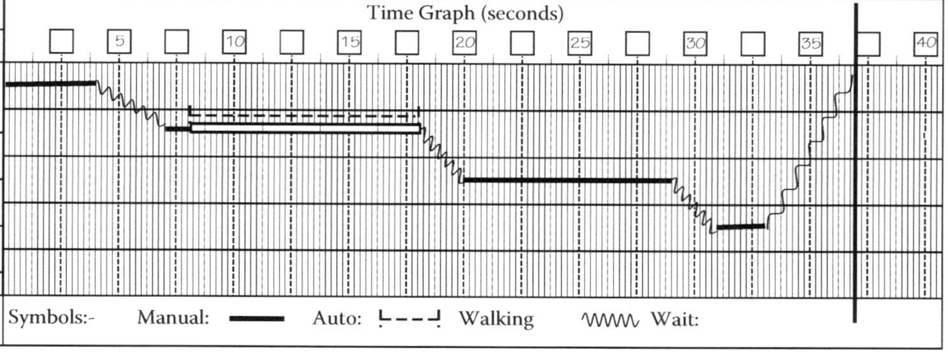

Symbols:- Manual: ▬▬▬ Auto: ⌐ ─ ─ ┘ Walking ∿∿∿∿ Wait:

FIGURE 8.8
Standardized Work Combination Table for T-shirt packing.

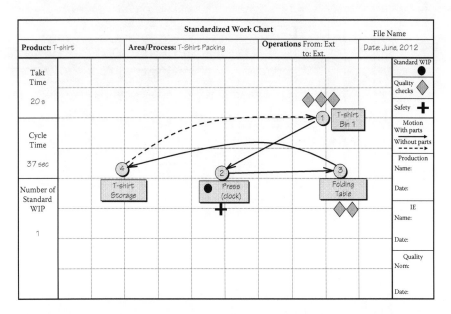

FIGURE 8.9
Standardized Work Chart for T-shirt packing—Team 1.

the second major step. Daniel advised that trainees should use red ink to represent the Takt time line on the combination table. He advised the use of green for the cycle time line.

Drawing the Standardized Work Chart presented a relatively low level of difficulty (Figure 8.9). However, Daniel needed to specify that the layout had to be scaled. He commented, "In a real situation, an AutoCAD file layout from which you remove all hindering details provides a good and neat layout to start with."

The challenge came from the Operator Work Instructions (Figure 8.10). The group needed to find the right drawing and picture to accompany the Key Points they had already defined in the preparation phase. Some explanations were needed. Therefore, Daniel spent some time explaining that a supporting picture should be clear enough for anybody to be able to understand without speaking the language used in the form. "As always, 'a picture is worth a thousand words,' and the rule applies here as well. The general rule is that you should include as many pictures or drawings as needed. However, one picture per Key Point should be enough." On the pictures, he also noticed that some groups were drawing the picture for unacceptable situations with variant ways of showing it: red-crossed or

Writing the Forms • 81

Home Appliance Inc.	Operator Work Instruction			Version 00	Plant:		Team 1	Document No.:	1
								Rev. Number	1
Reference:		Title:			Area	T-shirt packing	Mch		Page: 1/1
No.	Major Steps		✚ = Safety	◈ = Quality	◉ = Tip	Temps:	Drawings, Photos, etc.		
1	Pick up and inspect T-shirt		◈ – Sleeve and garment hems must be flat and wide enough to prevent curling. ◈ – The neckline should rest flat against the body ◈ – The neckline should recover properly after being slightly stretched				[1] Neckband — Bond on sleeve hem — Garment hem		
2	Load, start and unload press		✚ Maintain your hands out of the press while it is closing.				[2] Start button — Keep your hands in the hachured area		
3	Fold and inspect T-shirt		◈ Inspect the overall quality of the folding. ◈ Check that the logo is visible on the front side of the t-shirt				[3] Logo		
4	Store T-shirt								
Author Name: Function:		Signature/ Date	Verification Quality: HSE:	Signatures/ Dates	Approval Name: Group Leader	Signature/ Date	Operator Function:	Signature/ Date	Comment

FIGURE 8.10
Operator Work Instructions for T-shirt packing—Team 1.

barred with a "NOT OK." Once again, Daniel took the opportunity to restate his previous advice on Key Point description, "As for Key Point description, avoid unacceptable pictures. Focus on getting the Key Points right. Thereafter, concentrate on giving the right training."

Thomas sensed that the current module was heading toward the end. As always, before it ended he wanted to offer "a sort of summary," as he said. Daniel, now used to Thomas's so-called ritual, moved aside a little to provide more space in front of the chart for Thomas. Thomas shared a matrix of Elements of Standardized Work versus Standardized Work forms that displayed on which forms each element of STW could be found (Figure 8.11). This matrix provided unexpected insight into the subject. It appeared that the Standardized Work Chart was the one document that held all four elements of Standardized Work, visibly the most informative document of all. The Takt time and the Job Sequence emerged as the most shared elements, appearing in three of them. Daniel found the matrix informative, although he was not completely sure of the conclusion that could be derived from it.

"There ends the classroom part of the training. The next steps will be conducted on the shop floor," Daniel concluded … or wanted to conclude.

	Process Analysis Chart	STW Combination Table	STW Chart	Operator Work Instruction
Takt time	×	×	×	
Job sequence		×	×	×
Standardized Work in Process			×	
Key Points			×	

FIGURE 8.11
Crossing the elements of Standardized Work and the forms.

> *In defining a Key Point, the key is about what to do in order to be successful, not about what not to do. Because there are too many other ways of "what not to do" and only one way of "what to do," please focus on how to find the best, simplest way to describe it as well as the right drawing or picture to accompany the description.*

9
Shop Floor Application

It was time to move to the shop floor to carry on the training. The group took a little time to put on their Personal Protection Equipment (PPE). A 16-person group was too large for a workstation; therefore, Daniel proposed the group be split into two teams. They also identified two workstations that were visibly very similar. Each team would observe one workstation where it would apply learning received in the classroom. Everyone was now ready to head to the shop floor.

Daniel summarized the task given to the group in one word: "observe." He asked everyone to observe the way the operator was working for an hour. Daniel asked two participants to videotape the operator activity. He argued that this would help support discussions and fact-check when the group returned to the room. The tasks at hand when they returned were (1) agreeing on detailed steps and (2) grouping the detailed steps into measurement steps and defining measurement points in parallel.

An hour-long observation was too much for some people. Daniel noticed that after having observed a dozen cycles, some trainees' attention fluttered and they let themselves get distracted by other things. Some of them even complained that they had seen it all before; therefore, they felt like they were just wasting their time standing in front of the machine. Obviously, standing still somewhere to observe was very painful to some people. Daniel felt compelled to explain that observing was the key of the Standardized Work implementation. He showed some understanding by admitting that they would not be able to observe in the most efficient way from the beginning. "To the uninitiated, observation sounds like something very simple and even boring, but in reality, you need to train your eyes to be able to see. If not, you will just be like a myopic person trying to see without the right glasses."

84 • *Implementing Standardized Work*

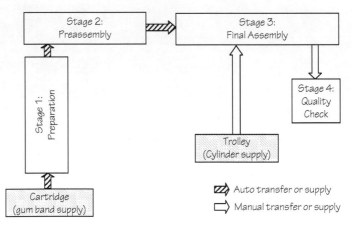

FIGURE 9.1
Flow and the layout of the assembly machine.

Before the work started, Steve, the IE manager, offered to share a poster depicting the machine flow and the layout with his colleagues to make sure everyone understood the way the assembly machine worked (Figure 9.1). He explained:

> The process is done in four stages. In the first stage, a gum band is pulled from a cartridge automatically and prepared automatically as well. The operator task on this stage is to flip the gum band. After preparation, the band is transferred to the second stage automatically. The operator's task then consists of completing a short assembly operation to fix a plastic part to the gum band. The rest of the operation is automated. When the operation in stage 2 is completed, the preassembled part is transferred automatically to stage 3. In this stage, the preassembled part is assembled with a cylinder. Everything is automated except the manual supply of the previously mentioned cylinder and the removal of the finished part. After this stage, the part is transferred manually to the fourth and last stage, where the operator performs quality checks.

Also to explain, Steve shared the process sequence including some time numbers with the whole group (Figure 9.2).

Steve offered to facilitate the calculation of the Takt time as well, and Daniel accepted. The customer demand, Steve explained, was 1000 parts per day. The work organization included a 5-minute communication break for communication to operators (called Top 5), two 25-minute breaks, and

Process Sequence

| Stage 1 | → | Stage 2 | → | Stage 3 | → | Stage 4 |

	Stage 1	Stage 2	Stage 3	Stage 4
Auto time	16"	36"	54"	0"
Man time	3"	4"	4"	7"

Stage 1: Preparation of gum band
Stage 2: Preassembly gum band and plastic components
Stage 3: Final assembly (assembly of cylinder and finishing)
Stage 4: Quality check and finish good removal

FIGURE 9.2
Process sequence of the assembly workstation.

a 45-minute lunch. This made a total of 100 minutes. Based on those data, the group calculated a Takt time of 68 seconds. This concluded Steve's technical presentation of the workstation. The group then moved to the shop floor.

After an hour or so spent on the shop floor, the team returned to the classroom, where they tried to follow the steps learned previously on the T-shirt packing process. Listing the detailed steps would be quick, at least that is what most of them thought before facing the situation. Problems came from the fact that the operator they observed had not been working the same way all the time. Most of the changes in the method were triggered by external disruptions: a breakdown here, material shortage there. Every time, the operator would try to accommodate to the change thereby creating even more disruptions. Because the method changed from cycle to cycle, the group was confronted with the classic question that very often arises at the beginning of Standardized Work deployment: what job sequence and target cycle time should be taken to start the process? This situation was also a teaching opportunity that helped them see firsthand why Daniel insisted that basic minimum process stability was needed before starting the Standardized Work deployment. Daniel, who had seen such situations before, let the group discuss, and because he knew that this could go on for hours he was ready to intervene. This was just what he did:

> Guys, I can see that you are struggling a bit to decide what the job sequence of the Standardized Work should be. You could carry on such a discussion for hours. In one of my first trainings, the group spent seven hours trying to figure out the right job sequence.

Most of the trainees were stunned by this information. Thomas then reacted, "Okay Daniel what should we do?"

"Well what would you suggest?" Daniel asked the attendees. He received a few suggestions, one of them being that they return to the shop floor and ask the operator to work as he or she is supposed to, and let the group observe. As Eric, the Excellence System manager, put it, "How could that be possible when some of those guys do not even know what the 'supposed to' way is? No one is trained or aware of any standard. Most of them have been taught by colleagues and from then have added their own improvements."

Daniel thanked Eric, and confirmed that he was completely right. Then he continued,

> As you all remember, you should reduce special cause variation impact on the process at the very beginning. However, you will never get the 100% stability as we said before. As you have seen on the shop floor, the operator will still sustain some variation, to which he may respond with self-directed variation as we just noted on the shop floor. This is a never-ending story. What do we do?

Daniel looked around but no answer came, and then he proceeded:

> Well, you should remember that your first set of Standardized Work forms will have a short life. They are not the most important documents. The ones that are important are those coming out of the improvement step. Again, because the first set of forms is ephemeral, you should not waste too much time debating the right job sequence. Adopt one, the most frequent, and use it to start your Standardized Work deployment process.

After this exchange, the group was ready to move to the next step. They were able to identify a sequence of cycles that were repetitive enough to apply the measurement method. First, they defined the measurement steps, performed the time collection, and calculated the modes as they had learned. Then, based on those data, they drew a Yama chart (Figure 9.3), the Standardized Work Chart (Figure 9.4), the Standardized Work Combination Table (Figure 9.5), and the Operator Work Instructions (Figure 9.6).

Daniel commented on the Key Points on the Standardized Work Chart. He noted that this time the team had depicted the three types of Key Points: Safety, Quality, Tips. He praised the group for the usage of short sentences to describe Key Points.

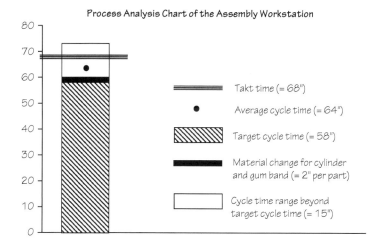

FIGURE 9.3
Process analysis chart (Yamazumihyo) of the assembly workstation.

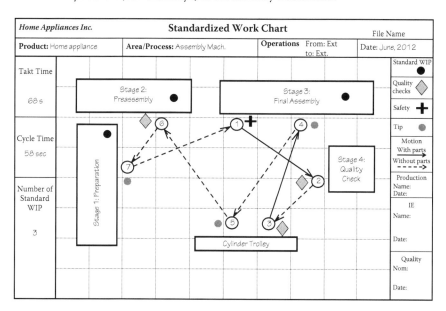

FIGURE 9.4
Standardized Work Chart of the assembly workstation.

Regarding the Standardized Work Combination Table, for the sake of simplicity, they decided to represent only the Stage 3 modules of the machine in the auto time, which was clearly the bottleneck. The other two auto modules were very far from being bottlenecks and therefore had little impact on the combination of worker–machine work. As a

Home Appliances Inc.

Standardized Work Combination Table

Title: Assembly Mach

By:

Takt Time: 68
Volume: 1000 Parts/day
Date: June, 2012

Comments

N°	Major Steps	Time				Time Graph (seconds)
		Manu	Auto	Wait	Walk	
1	Unload finished part and start mach.	2	28		3	
2	Check finished part.	7		6	1	
3	Pick up the cylinder	2		3	3	
4	Load cylinder and start mach.	2	26		3	
5	Prepare the cylinder	3			3	
6	Fix plastic part	4			1	
7	Flip the gum band	3		10	2	
	Total	23	54	19	16	

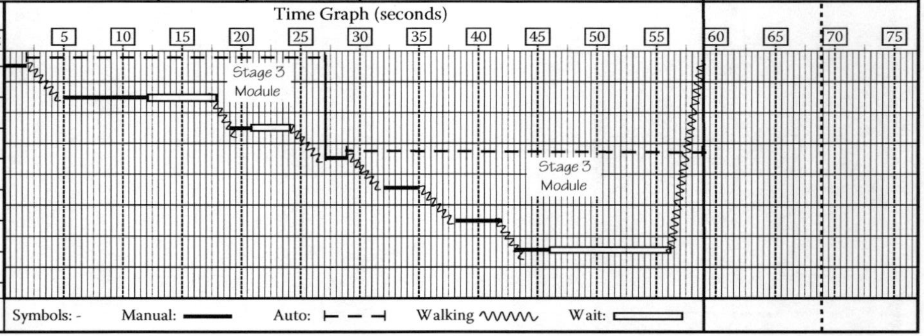

Symbols: -   Manual: ———   Auto: ├ ─ ─ ┤   Walking: ∿∿∿   Wait: ▭

FIGURE 9.5
Standardized Work Combination Table of the assembly workstation.

Home Appliance Inc.	**Operator Work Instructions**	Version 00	Plant:	Document No.: 1 / Rev. Number 1
Reference:	Title:	Area	Mch	Assy Workstation — Page: **1/1**

No.	Major Steps	✚ = Safety / ◈ = Quality / ● = Tip		Pictures, drawings, etc.
1	Unload finished part and start mach. (Stage 2 & 3)	✚	Left hand on the upper part of the machine and right hang on the part	1
2	Check finished part	◈	Special inspection of Gum–plastic junction: circle the three screws with black marker to confirm presence and tightness are checked	2
3	Pick up the cylinder	◈	Check the reference using the color code	3
4	Load cylinder and start stage 3	●	Load cylinder horizontally	4
5	Prepare the cylinder	●	Pull plastic protection simultaneously with two hands.	5
6	Fix plastic part	◈	Keep the plastic part horizontal when clipping	6
7	Flip the gum band	●	Grasp the band at 1/4 from each bottom	7

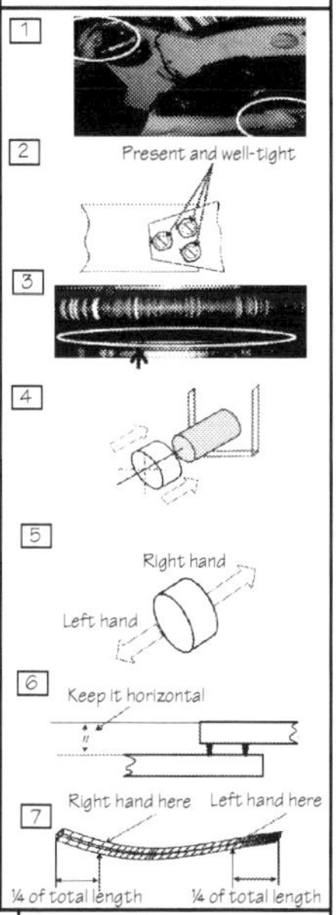

Author	Signature/ Date	Verification	Signature/ Dates	Approval	Signature/ Date	Operator	Signature/ Date	Comment
Name :		Quality :		Name :				
		HSE :						
Function:		IE :		Group Leader		Function:		

FIGURE 9.6
Operator Work Instructions of the assembly workstation.

consequence, they opted not to represent them on the paper to make the diagram less encumbered.

Writing Operator Work Instructions was an excellent opportunity to practice Key Point identification. It was also "a welcome occasion to practice how to find the right picture or come up with the right drawing to convey the just-needed message," as Daniel stated (Figure 9.6).

The day was coming to an end when the group completed its work. As Daniel commented,

> This has been a very productive day and I congratulate you for your hard work. I hope the application of learning on the shop floor workstation as we just did gave you an even better understanding of how to analyze and write Standardized Work forms. Before you go back to enjoy a well-deserved rest with your family, I would like to share some Key Points regarding Standardized Work forms with you. Well, actually, I prefer that Thomas

does it. Thomas, I know that this is your favorite exercise. Could you please facilitate the summary of today's work?

> *Observation is the key of the Standardized Work implementation. To the uninitiated, observation sounds like something very simple and even boring, but in reality, you need to train your eyes to be able to see. If not, you will just be like a myopic person trying to see without the right glasses.*

10

Takeaway

Thomas first expressed a lot of gratitude to Daniel for what he called "an excellent day of learning." Then he continued while drawing on the flipchart:

> We will not be very long. Here are my main takeaways. First, the conditions for implementation of the Standardized Work: The process must have minimal variation and be repetitive ... ideally, a cyclic operation (Figure 10.1). The Standardized Work, at least in the format you taught us, has four elements: Takt time, Job Sequence, Standardized Work in Process, and Key Points. You added Key Points because you believe that they are the "knowledge reservoir" and therefore belong as elements of Standardized Work. On the forms, you taught us how to write four documents: the Process Analysis chart or Yamazumihyo, as some Japanese companies call it, the Standardized Work Chart, the Standardized Work Combination Table, and the Operator Work Instructions. This list of forms is similar to Toyota's, but you have removed the Process Capacity Sheet and included both the Process Analysis Chart and the Operator Work Instructions.
>
> We have also learned the importance of focusing on the method versus the time while implementing Standardized Work. You underscored that the Standardized Work approach is built on the notion that the "right method will lead to the right time." Therefore, we should focus on the method to achieve highest productivity (Figure 10.2).

Thomas took a long breath and then proceeded, "Last but not least: the steps to writing Standardized Work. We have practiced that twice. There are three steps, which look very simple and logical: data collection, preparing for writing, and writing itself (Figure 10.3)."

92 • *Implementing Standardized Work*

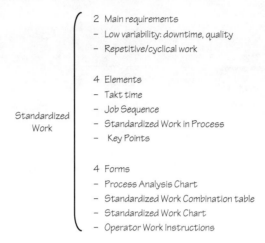

FIGURE 10.1
Main requirements, elements, and forms of Standardized Work.

Classical Approach to Standard	Standardized Work Approach
Description: "focus on the What and enforce"	Description: "focus on the How and build together"
Here is your time objective based on IE calculations.	Let us find together the best way to perform this operation.
Please manage yourself to reach it.	Everyone is trained to the resulting best method (in terms of safety, quality, and productivity).
… And by the way to make sure that you achieve it, your wage* will depend on how you perform vs. the time.	To make sure you are following this best method, auditing will be carried out to help identify problems and solve them together.
Consequence: The focus is on the time that people do not always know how to reach. It creates conflicts. Also workers take risks in their task to achieve the time objective, which hampers safety and quality.	Consequence: The focus is on the method. Workers contribute, share, and create team work: motivation and mastery of the job increases.
Bottom Line: Safety, Quality, Productivity, and Morale go down.	Bottom Line: Safety, Quality, Productivity, and Morale go up.

*This could be managed with individual worker (piece-work) or with a group of workers.

FIGURE 10.2
Classical approach versus Standardized Work approach.

It was getting late; Daniel did not want to wait any longer to release the group, which looked somewhat tired. He greeted them and concluded, "See you tomorrow for the next module of the training. We will be focusing on process improvement.* Have a nice evening."

---

* Process improvement is the subject of a subsequent book in the same series.

> 1. **Data collection**
>   1. Identifying detailed steps
>   2. Defining measurement step and measurement points
>   3. Collecting data
> 2. **Preparing the elements of Standardized Work**
>   1. Takt time
>   2. SWIP
>   3. Job sequence
>   4. Key Points
> 3. **Writing the forms**
>   1. Process analysis (Yama)
>   2. Standardized Work Chart
>   3. Standardized Work Combination Table
>   4. Operator Work Instructions

FIGURE 10.3
Steps to writing Standardized Work forms.

The room was emptying when Thomas approached Daniel to thank him again for the training. They went on to talk about the whole deployment of Standardized Work. Thomas talked about the huge challenge they were headed for. "It certainly won't be a cake walk," he commented. Daniel seized this opportunity to encourage him:

> Thomas, you have inherited an excruciatingly difficult situation. You have no choice but to "shoot higher than you know you can do" like William Faulkner said, you will need to be "better than yourself." I can see that this is just what you are trying to do through this very ambitious deployment of Standardized Work. This is not a sprint but a marathon. You will have good days and tough days, but if you stay the course, I have no doubt that success is within your reach.

Index

A

absenteeism, 50
accumulation, unfinished parts, 59
accuracy *vs.* practicality, 39
activity, *see* Operators
Africa, *see* EMEA (Europe, Middle East, and Africa)
approximations, 38–39
auditing, 7
AutoCAD, 80
auto times, 78
auxiliary operations
    periodical, 26, 29
    T-shirt simulation, 28
averages, 44

B

baking example, 61
bottlenecks, 68–69
buffer time example, 6

C

cake example, 61
capability, satisfying customer demand
    current standards, 14, 15
    overview, 12
capacity analysis bar, *see* Yamazumihyo
capacity of process
    Process Analysis, 1
    Yama tool, 7, 68–71
change procedures, 26
changes, Key Point stability, 57
cities analogy, 23
classical approach, 92
cleaning procedures, 26
code change per part, 72, 74
combination table, *see* Standardized Work Combination Table
complex operations, 27
computations make sense, 77, 78
cooking example, 61
copying others, 50–51, 52, 53
core operations
    reports, 4
    T-shirt simulation, 28
counting steps *vs.* time measurement, 37–39
current state, capturing
    Process Analysis, 1
    Standardized Work deployment, 7
customer demand, *see also* Capability, satisfying customer demand
    Process Analysis, 1
    simulation, 25–29
cycle time of operations
    beyond target, 72, 74, 75
    current standard, 13
    green ink, 80
    overview, 12
    target cycle time, 71, 72, 73
    Yama chart, 72, 73
Cycle Time Range, 72, 75–77
cycle time reduction
    current standard, 14, 15
    overview, 12–13
cyclic operations
    prioritization, 28–29
    repetitiveness, 26, 86
    variation by operators, 85

D

data, existing documents, 22–23
data collection, *see also* Time collection
    group formation, 31
    job instructions, 34
    measurement steps and measurement points, 36–41
    overview, 31, 93
    performing, 42–47

steps, 35–36, 91
team organization, 32–34
death spiral, 25
debating job sequence, 85–86, *see also* Scholarly arguments
Deming, W. Edwards, 52
demotion, 4
deployment priorities, 29
Design of Experiments (DOE), 63
details
    bogged down in, 38–39
    step descriptions, 35–36
deterministic causes, 1
discrepancies of results, 40
discussions, endless, 56, 58, 85–86
disruptions, 42, 85
documents, existing, 22–23
DOE, *see* Design of Experiments (DOE)
drama, IE personnel, 39
drawings, 80–81, 82, 89
dry run, cycles, 41–42

E

economies of scale, 26–27
elements, preparation overview, 93, *see also specific element*
EMEA (Europe, Middle East, and Africa), 2
endless discussions, 56, 58, 85–86
equipment
    breakdown, 50
    shift challenges, 25
ergonomic issues, 45
estimation, stability, 1
Europe, *see* EMEA (Europe, Middle East, and Africa)
existing standards and documents, 9–23
expectations, Takt time, 2
external disruptions, 85
eyes, observation training, 83, 90

F

fairness, lack of
    Cycle Time Range, 77
    minimum repeatable value, 44
    standards development, 45–46
Faulkner, William, 93

four-step solution, 40–41
Fourth Generation Management: The New Business Consciousness, 61

G

Gemba, 4
green ink, cycle times, 80
group formation, 31, *see also* T-shirt simulation
"Guru" label, 3

H

hammer analogy, 68, 78
honesty, 77, 78
hotel room swipe card example, 61
human-centeredness, 26, 29
human disruptions, 50

I

ill-adapted tools, 68, 78
implementation, easy, 77, 78
Implementing Standardized Work: Measuring Operators' Performance
    classification of variation, 50, 71
    initial steps, 2
    Operators' Performance Measurement module, 5
    pitch of simulation, 25
    special cause variation, 71
    spiral, 49
    steps, Standardized Work deployment, 6
improvement opportunities
    important documents, 86
    measurement steps, 37
    overview, 7
    standards importance, 16
indispensability, undesirable, 3
Industrial Engineering standards
    forms, 9, 10–11
    Standardized Work forms comparison, 16–17, 19–20
    standard time breakdown, 45
inventory, hiding, 25
ironing press operation, 32, 34

J

Japanese words, 67
job instructions, 34, *see also* Operator Work Instructions
Job Sequence
  debating, 85–86
  as element of forms, 53, 58, 59
  minimum process stability, 85–86
  overview, 2
  preparation results, 65
Joiner, Brian, 61–62, 63

K

Key Points, *see also* Reviews
  accuracy *vs.* method, 39
  as element of forms, 53, 59–64
  number of, 63
  Operator Work Instructions, 80–81, 89
  overview, 2
  preparation results, 65
  short sentences, 86, 89
  stability, Takt time changes, 57
  Standardized Work deployment, 7
knowledge reservoir, 53

L

layout
  of machines for observation, 84
  scaled, 80
leadership, 35, 51–52
Lean, leading to demotivation, 4
literacy level, 22, 23
losses
  Process Analysis, 1
  Yama tool, 7

M

machines
  data, existing documents, 22–23
  poster, machine flow, 84
  stable processes, 49
  Takt time, 56
major steps, 58, *see also* Job Sequence
making sense, computations, 77, 78
manual times, 78
material change per part, 72, 74
matrix, writing the forms, 81–82
maximum values, 44
measurement of operators' performance, *see* Implementing Standardized Work: Measuring Operators' Performance
measurement steps and measurement points
  data collection, 36–41
  defined, 36–37
methods
  changes in, 50, 85
  focus on, 46–47
  leading to right time, 39, 40, 91
  vs. time accuracy, 39, 40
Method Time Measurement (MTM), 39
Middle East, *see* EMEA (Europe, Middle East, and Africa)
minimum values, 44
modes, 44
morale, 2, 4
multistage bar, *see* Yamazumihyo

N

needles analogy, 68, 78
negative presentation, 62, 80–81, 82
non-cyclic/non-repetitive operations, 26, 29
non-human-related disruptions, 50
"not okay" situations, *see* Positive presentation

O

observation, 83–86, 90
OEE, *see* Overall Equipment Efficiency (OEE)
Operation Instruction Sheet, 50
operators
  activity, 12
  changes in methods, 85
  current standards, 14, 15, 86
  Standardized Work form, 16, 19
  working to standard, 12, 13–14
operators' performance, *see* Implementing Standardized Work: Measuring Operators' Performance

Operator Work Instructions, *see also* Job instructions
   assembly workstation, 86, 89
   overview, 2
   as tool, 7, 50
   writing the form, 78, 80–81
order of deployment, priorities, 29
outliers, removing, 45
Overall Equipment Efficiency (OEE), 4, 49, 55
overcapacity, hiding, 25
overviews, 1–2, 91–93

P

pace rating, 45
packing operation, *see* T-shirt simulation
PDCA (Plan-Do-Check-Act), 52
performance rating, 45
periodical auxiliary operations
   prioritization, 29
   task standardization, selection, 26
personnel-related variation, 1
physical fatigue recovery, 45
pictures, 80–81, 82, 89
planning, *see* PDCA (Plan-Do-Check-Act)
popovers example, 61–62, 63
positive presentation, 62, 80–81, 82
poster, machine flow, 84
power, loss of, 4
practicality *vs.* accuracy, 39
preparation for training, 5
press-handler, simulation, 32
prioritization, 28–29
problem-solving
   copying others, 58
   helping others to perform, 3
Process Analysis, stability estimation, 1
Process Analysis Chart, *see* Yamazumihyo
Process Capacity Sheet
   replaced, 50
   Yama comparison, 68–70
process improvement
   overview, 7
   standards importance, 16
process sequence, observation help, 84, 85
product data, 22–23
productivity, Key Points, 60

Q

quality
   checks, 34
   Key Points, 60, 65, 86
"quick-and-dirty" standards development, 45
quiz on standards, 12–16, 20

R

random operations, 27
reading ease, 22, 23
red ink, Takt time, 80
repeatability, 1
repetitiveness
   by design, 50
   prioritization, 28–29
   reorganizing to create, 27
   task standardization, selection, 26–27
reports, 4
resources, balancing, 19
result discrepancies, 40
reviews
   goals, Standardized Work forms, 13
   Key Points, 54, 64
   main features, Standardized Work forms, 23
   observation, 90
   operator working to standard, 13
   shop floor application, 90
   task standardization selection, 29
   tools, choosing and adapting, 52

S

safety, Key Points, 60, 65, 86
scholarly arguments, 56, 58, *see also* Debating job sequence
screwing operation, 26–27
seconds
   counting steps *vs.* time measurement, 38
   Standardized Work form, 19, 21–22
   Takt time computation, 57
sensibility of computations, 77, 78
sequence, observation help, 84, 85
shift challenges, 25
shop floor application, 83–90

shutdown procedures, 26
sickness, 50
simplicity, 77, 78
simulation, *see* T-shirt simulation
size change per part, 72, 74
stability
    estimation by Process Analysis, 1
    minimal, 49, 85–86
standard, operators working to, 12, 13–14, *see also* Takt time
Standardized Work, *see also specific process*
    classical approach comparison, 92
    elements, 54–64, 91, 92
    forms, 92
    main requirements, 92
Standardized Work Chart
    assembly workstation, 86, 87
    overview, 1, 16, 17
    as tool, 7
    T-shirt simulation, 80
    writing the form, 78, 80
Standardized Work Combination Table
    challenges, 78
    overview, 1–2, 16, 18
    as tool, 7
    T-shirt simulation, 78, 79, 86–88
    writing the form, 78, 79
Standardized Work deployment
    chart of steps, 6
    Key Point, 7
    training, 7
Standardized Work forms
    Industrial Engineering standards comparison, 16–17, 19–20
    Job Sequence, 58
    Key Points, 59–64
    main features of, 22
    modifications, 50–51
    overview, 53–54
    Standardized Work in Process, 58–59
    Takt time, 54–58
Standardized Work in Process (SWIP)
    as element of forms, 53, 58–59
    overview, 2
    preparation results, 65
standards warm-up quiz, 12–16, 20
start-up procedures, 26
steps, data collection, 35–36

sterile discussions, 56, 58, *see also* Debating job sequence
stochastic variation, 1, 71
stopwatch keeper, simulation, 32
summary, 91–93
SWIP, *see* Standardized Work in Process (SWIP)

T

takeaway, 91–93
Takt time
    clear expectations, 22
    computing, 54–56
    as element of forms, 53, 54–58
    observation, 84–85
    overview, 2, 17, 19
    preparation results, 65
    red ink, 80
    Yama chart, 71, 72, 73, 76
target cycle time, 71, 72, 73
tasks, selecting for standardization, 25–29
teams, *see also* T-shirt simulation
    layout of position, 31–32
    organization, 32–34
"The Popovers," 61–62, 63
time and motion studies, 9–11, 39–40
time collection, *see also* Data collection
    sheet example, 32, 33
    Team 1, 43
    using, 42–47
    Yama chart, 72
time collector, simulation, 32
tips, 60, 62–63, 86, *see also* Key Points; Reviews
tools, *see also specific type*
    capturing current state, 7
    choosing and adapting, 50–51, 52
    well-adapted, 68, 78
Toyota, 50–51, 52
tradeoff, accuracy/practicality, 39
trade union agreements, 46
training
    adaptation for group size, 5–6
    eyes, 83, 90
    Key Points, 54
    overview, 3–7
    preparation for, 5

Standardized Work deployment, 7
trueness, 77, 78
trust, lack of, 2
T-shirt simulation
    assembly workstation, 86, 87
    challenges, 80–81
    data collection, 31–47
    Key Points, 62–65
    maximum machine capacity, 69
    observation, 83–86
    Operator Work Instructions, 80–81
    Process Capacity Sheet, 70
    process flow chart, packing operation, 59–60
    shop floor application, 83–90
    Standardized Work Chart, 80
    Standardized Work Combination Table, 78, 79
    task standardization selection, 25–29
    Yama chart, 71–77

U

understanding, simplicity of, 77, 78
unfinished part accumulation, 59

V

variability and variations
    absorbing or conquering, 6
    example, 6
    hidden cost, 7
    order to be addressed, 1
    Process Analysis, 1
    Yama chart, 71
voice of the customer, *see* Takt time

W

wait times
    current standard, 14, 15
    measurement steps, 37
    overview, 12
    Standardized Work forms, 22, 78, 80
walk times
    current standard, 14, 15
    measurement steps, 37
    overview, 12
    result discrepancies, 40
    Standardized Work forms, 22, 78, 80
warm-up quiz, 12–16, 20
waste, *see also specific form of waste*
    measurement steps, 37
    Takt time, 55
well-adapted tools, 68, 78
work sequence, *see* Job Sequence
writing the forms
    matrix, 81–82
    Operator Work Instructions, 78, 80–81
    Standardized Work Chart, 78, 80
    Standardized Work Combination Table, 78, 79
    steps overview, 91, 93
    Yamazumihyo, 67–78

Y

Yamazumihyo
    assembly workstation, 86, 87
    defined, 1
    steps for building, 72–74
    as tool, 7
    writing the form, 67–78

About the Author

Alain Patchong is the Director of Assembly at Faurecia Automotive Seating, France. He also holds the title of Master Expert in Assembly processes. He was previously the Industrial Engineering Manager for Europe, the Middle East, and Africa at Goodyear in Luxembourg. In this position, he developed training materials and led a successful initiative for the deployment of Standardized Work in several Goodyear plants.

Before joining Goodyear, he worked with PSA Peugeot Citroën for 12 years where he developed and implemented methods for manufacturing systems engineering and production line improvement. He also led Lean implementations within PSA weld factories.

He teaches at Ecole Centrale Paris and Ecole Supérieure d'Electricité, two French engineering schools. He was a finalist of the INFORMS'[*] Edelman Competition in 2002 and a Visiting Scholar at MIT[†] in 2004. He is the author of several articles published in renowned journals. His work has been used in engineering and business school courses around the world.